1295

California Natural History Guides: 33

GRASSES IN CALIFORNIA

BY

BEECHER CRAMPTON

UNIVERSITY OF CALIFORNIA PRESS

BERKELEY • LOS ANGELES • LONDON

CALIFORNIA NATURAL HISTORY GUIDES
Arthur C. Smith, General Editor

ACKNOWLEDGMENTS

By permission of the trustees of the Jepson Herbarium, the following figures are used from W. L. Jepson, *A Manual of the Flowering Plants of California*, 1970: 9, 12, 16, 21, 24, 26, 37, 39, 40, 42, 52, 56, 60, 61, 64, 68, 71, 73, 74, 81, 87.

All other figures were drawn, and all color pictures were taken, by the author.

Cover photographs
Upper left: *Melicia torreyana*, Torrey Melic.
Upper right: *Sporobolus airoides,* Alkali Sacaton.
Lower left: *Stipa pulchra*, Purple Stipa.
Lower right: *Sitanion jubatum*, Big Squirreltail.

University of California Press
Berkeley and Los Angeles, California

University of California Press, Ltd.
London, England

© 1974 by The Regents of the University of California

ISBN: 0-520-02507-5 (alk. paper)

Library of Congress Catalog Card Number: 73-80829

Printed in the United States of America

3 4 5 6 7 8 9

The paper used in this publication is both acid-free and totally chlorine-free (TCF). It meets the minimum requirements of American Standard for Information Sciences—Permanence of Paper for Printed Library Materials, ANSI Z39.48-1984. ♾

CONTENTS

INTRODUCTION	5
THE GRASS PLANT	6
Perennials	6
Annuals	10
PARTS OF THE GRASS PLANT	11
Stems	12
Leaves	12
Flowering Structures	13
Spikelets	15
Epidermal Structures	19
Seed Dispersal	20
USES OF GRASSES	22
IDENTIFICATION OF GRASSES	24
RELATIONSHIPS IN GRASSES	24
DISTRIBUTION OF GRASSES IN CALIFORNIA	29
The Valley-Foothill Area	29
The Montane Area	33
The Coastal Area	35
The Desert Area	36
A KEY TO THE GRASS GENERA	39
DESCRIPTION OF THE GENERA AND SPECIES	49
THE COLLECTION OF GRASSES AND THE PRESERVATION OF SPECIMENS	163
GLOSSARY	167
SELECTED READINGS AND REFERENCES	171
INDEX	172

INTRODUCTION

Grasses are the most abundant, widespread, and useful plants in California. The cereal grasses—wheat, barley, rice, oats, and corn—are planted on thousands of acres in the state. Extensive areas are covered with natural grasses, and these grasslands support large herds of domestic as well as wild animals. Some grasses are valuable as turf, and others are used as ornamentals.

The varied climates of California favor development of many kinds of grasses, especially species of typical temperate zone origin, that is, those that grow during the cool season of the year. Other California species are of subtropical origin and develop during the warm or hot part of the year. At low elevations in the state these warm-season grasses grow where there is summer irrigation and are primarily weeds of croplands or ornamental plantings.

Sixty-five genera and 162 species of native and introduced grasses are described in this guide, including the important crop grasses. Most of the species are the common grasses, but some were selected for an outstanding character or use. Most of the 162 species are widely distributed, important as animal forage or crops, planted as ornamentals, or used to check soil erosion. It was impossible to include all the grasses that grow in California, but the ordinary and most useful ones are described.

Many of these grasses might be encountered on a walk or car trip through the valleys or the foothills during the spring, while others might be found as weeds in or about cities and towns. Grasses spread to the far reaches of the mountains and to the desert—wherever you are so are the grasses. Familiarity with the various species presented here may stimulate a further exploration into this immensely important and quite fascinating group of plants.

THE GRASS PLANT

Grasses are primarily herbaceous, which means that except for the bamboos and certain tall reedlike grasses, they do not develop woody stems. The grasses growing naturally in California eventually die back to the ground at the approach of the hot, dry summer in the lowlands, or the cold of winter, or the first snowfall in the mountains. Some grasses renew growth yearly from the previous seasons crown, or from creeping stems either above or below ground level—these are *perennials*. Other grasses complete their entire growth and reproductive cycle within one season or year's time and die at the end of this period—these are the *annuals*, which depend solely upon the production and dispersal of seed for their perpetuation.

The perennial or annual growth characteristics of grasses are extremely important in determining their use as crop plants, forage plants, or as ornamentals. Knowing whether a grass is annual or perennial greatly facilitates identifying the species.

PERENNIALS

Under favorable, or at least ordinary environmental conditions, perennial grasses emerge from dormancy or revive by active vegetative growth about the base of the plant. (pl. 1a). A canopy of fresh leaves is soon produced among the weathered, grayish or blackish stems and leaves of the previous season's growth. The new shoots, or *tillers*, develop from buds produced at the base of these year-old stems. The young shoots remain vegetative during the entire growing season, or some may elongate and produce an *inflorescence* or flowers at the tip. The presence of both sterile and fertile shoots in the same clump further distinguishes them from annuals. Perennial grasses are of two kinds—*bunchgrasses* and *sod-forming grasses*.

Bunchgrasses.—These are densely tufted with leaves mostly at the base and erect or somewhat spreading *culms* or stems (pl. 1b). Most bunchgrasses are adapted to dry soils and constitute important forage elements of semi-arid and arid rangelands (pl. 1c). Certain individual bunchgrasses may live extremely long, as much as 100 years. The center usually dies out leaving an outside ring of green and active growth (pl. 1d). Further die-out breaks up the ring and isolates tufts of the grass and, in essence, is a type of vegetative reproduction. Some bunchgrasses, such as Dallisgrass (*Paspalum dilatatum*) and Timothy (*Phleum pratense*), prefer soils that are moist during most or all of the growing season, and they are valuable as forage in irrigated pastures or meadows.

Bunchgrasses usually produce an abundance of seed. Though the tufted parent plant increases its size from year to year, it is seed that ultimately maintains the species in a given area and allows it to spread elsewhere. Rangeland with good stands of bunchgrasses must be carefully managed to conserve the existing parent plants and at the same time allow for adequate production and seedling development.

Commonly the tufted grasses develop a strong, fibrous root system that deeply penetrates the soil. This intricate network of root fibers primarily supplies water and minerals to the growing grass plant, but secondarily the roots serve to hold soil particles firmly in place and reduce soil erosion. If you pull out a bunchgrass plant and observe the amount of soil among the roots, you will be aware of its soil-holding capacity. Any dense grass cover greatly reduces the erosional effects of wind and water while providing a pleasing landscape.

Sod-forming grasses.—These produce erect flowering stems and also creeping stems, which extend either above or below the ground level or sometimes both (fig. 1), and which allow the plant to spread rapidly in any direction. The green and leafy stems that are ordinarily prostrate upon the ground, and which can root at some or all of the nodes, are called *stolons* (fig. 1). Stolons are sometimes arched and root only at those nodes in contact with the soil surface. Each node of a stolon that roots in the soil is a potential new plant and will produce a tuft of foliage, eventually some erect stems, and finally a well-developed crown, which then renews production of stolons.

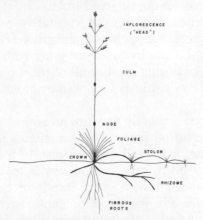

Fig. 1. The grass plant

The stems which penetrate the soil and commonly form an underground branching network are called *rhizomes* (fig. 1). They are whitish, yellowish or brownish in color, and their leaves are either reduced to sheaths, or, if the blades are developed, they are tough, and sharp-pointed. Each node of a rhizome is capable of rooting or branching, though commonly this does not occur. Rhizomes which occur near the surface of the soil produce a tuft of green foliage and aerial stems, develop a crown, and resume production of underground stems. Eventually this forms a rather uniform cover of foliage and stems (pl. 1e), in contrast to the isolated and often widely spaced condition of the bunchgrasses (pl. 1c, g). Plants with combination of rhizomes and stolons, as in Bermudagrass (*Cynodon dactylon*) and Kikuyugrass (*Pennisetum clandestinum*), form a very dense sod and are extremely long-lived and tenacious. Probably the oldest living things in the world are not the redwoods or bristle-cone pines but such lowly grasses as these that live indefinitely in their native habitat by continual vegetative reproduction.

A sod-forming grass, such as Kentucky Bluegrass (*Poa pratensis*), is usually quick to recover from excessive grazing or mowing. New, erect stems arise from the crown of the plant as well as from the nodes of the rhizomes. Such grasses are also able to withstand considerable trampling by livestock without serious damage to the plant, since the rhizomes are relatively undisturbed by animal hooves.

The strong tendency for vegetative recovery and rapid spread by vegetative means has, in many sod-forming grasses, greatly reduced the seed-producing ability. Although they may grow erect flowering stems, the seed set is often absent or at best very poor. Perennials are largely cross-pollinated even though there are both male and female parts in each flower. Sometimes there has been a gradual abortion of one sex in all of the flowers of a single parent plant of a sod-forming grass. As a result a whole colony of vegetatively produced plants can be unisexual with no opportunity to produce seed. Saltgrass (*Distichlis spicata*), a low and gregarious species, often produces such colonies. In some areas there may be intermingling of both staminate and pistillate colonies with resultant seed production. But not all sod-forming grasses have poor seed production. Kentucky Bluegrass, as an example, is consistently high in seed production allowing the plant to spread actively by both vegetative and sexual means.

Sod-forming grasses ordinarily are best developed on

soils having adequate moisture during the growing season. Certain meadow species in the mountains are rhizomatous, and those used as lawns or in irrigated patures are often so. Rhizomatous or stoloniferous species seldom persist for very long on dry soils, however Big Galletagrass in the arid Mojave and Colorado deserts is certainly a notable exception.

The perennial seedling is far less aggressive than the annual. Ordinarily a perennial increases its diameter during the first growing season and some flowering culms are occasionally produced, but usually they don't appear until the second growing season. In competition with annuals perennial seedlings often fare badly, especially on semi-arid or arid soils where competion among perennial seedlings for water is often critical, and many die.

ANNUALS

Annual grasses complete a life cycle within a single season or at most a year's time. (pl. 1f). Following germination of the annual grass seed there is a period of rapid vegetative growth; however, unusually cold temperatures or extended drought periods may retard this process. Annual seedlings under usual environmental conditions are aggressive, quick growing, and soon produce flowers. In contrast to the perennials, every shoot of the annuals is fertile and terminates in a flowering structure, since seed is essential for perpetuation of the species. A further guarantee of seed production is the self-fertility of annual grass flowers. Soon after emergence of the inflorescence, the foliage begins to wither about the base. The leaves gradually die on the stem as flowering, seed set, and finally maturation of seed proceeds. At maturity the entire plant dies and all traces of chlorophyll vanish (pl. 1f). Eventually seeds are deposited upon the ground and these become the source of the next year's crop of plants. The rapid spread of annuals is the result of their great numbers, large seed crops, and competitive ability.

At lower elevations in California, excluding deserts, annual grasses are the most important plants for animal forage. In the spring they literally form a green carpet over the valleys and foothills. The early winter and spring rains along with a cool growing season favor development of such grass genera as *Avena, Bromus, Hordeum, Festuca* and *Lolium*. Some or all species in these genera, because of their time of germination and active vegetative growth, are called *winter annuals*. Cattle and sheep graze the winter annuals during the early development when the protein levels in the plants are highest. Some annuals, such as Soft Chess (*Bromus mollis*), are nutritious even after seed maturation because of nutrients stored in the seed and the retention of the florets.

The onset of the hot, rainless summer and the exhaustion of soil moisture causes the winter annuals to die. On summer-irrigated lands of the valleys a different series of annual grasses then develop. These are the *summer annuals* and belong to genera with tropical or subtropical affinities, such as *Echinochloa, Panicum, Eragrostis, Chloris, Leptochloa* and others, which are common weedy species of croplands, pastures, and ornamental plantings. Others, such as *Oryza, Sorghum* and *Zea* are raised in large acreages for their grain. The first freezing temperatures of fall or winter effectively kill out these summer annuals or warm-season grasses, but not before large quantities of seed are shed upon the ground or are harvested, thereby ensuring the crop for the next summer season.

PARTS OF THE GRASS PLANT

A careful examination of all parts of a grass plant is necessary to distinguish one species from another. For each part or structure reference points must be established and botanical terminology used for each to facilitate grass description and identification.

STEMS

The grass stem that supports the leaves and elevates the flowering structure is called the *culm* (fig. 1). In most grasses the culm is cylindrical, but sometimes it is more or less flattened. It is jointed with solid *nodes* and, in many grasses, hollow *internodes*. Culms are usually strictly erect or they may ascend from a curved or sharply bent (geniculate) base. Some stems lie upon the ground for a distance and then curve upwards at their tips, which condition is called *decumbent*. Sometimes the bases of the culms are markedly swollen or bulbous as in some *Melica* species, Timothy (*Phleum pratense*) and Hardinggrass (*Phalaris tuberosa* var. *stenoptera*). Culms ordinarily branch about the base producing essentially undivided simple stems. However many of the summer flowering grasses of low elevations branch from nodes well above the base. Such nodal branching produces a somewhat bushy and commonly top-heavy structure with many more seeds, since each branch terminates in an inflorescence.

LEAVES

Grass leaves are alternately arranged on opposite sides of the stem and are thereby 2-ranked or *distichous*. The lower portion of the leaf usually closely surrounds the stem and is called the *sheath* (fig. 2). Sheaths usually have free margins and hence are referred to as *split*. In some genera such as *Melica, Glyceria* and *Bromus,* the margins are fused or *closed,* forming a tube about the stem.

The upper portion of the leaf is called the *blade* and is divergent from the stem. It may be *flat*, or the margins may be *rolled* in upon each other. Tightly rolled blades, as in some *Stipa* or *Festuca* species, are quite slender and almost hairlike in width (filiform). Rarely some blades may be folded or V-shaped in cross-section as in Orchardgrass (*Dactylis glomerata*). Blades are usually strongly nerved on the upper surface, the nerves commonly elevated as *ribs*. At the junction of the sheath and blade is the *collar*, usually a lighter-colored area, although sometimes dark-colored, and sometimes with tufts of hair on either side of the stem, or with the margins laterally extended as small lobes or *auricles*.

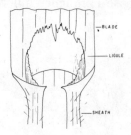

Fig. 2. The grass leaf

Inside the juncture of the sheath and blade is the *ligule* (fig. 2). It is an extension of the inner lining of the sheath and is often a thin, usually whitish membrane or, sometimes, it may be a fringe of hairs. Occasionally the ligule is absent, as in Watergrass (*Echinochloa crusgalli*). Its length or shape is often important in species determinations, and whether it is a membrane or a fringe of hairs is of value at the generic or tribal level.

FLOWERING STRUCTURES

The flowering structure, or *inflorescence*, is elevated by, and borne at the end of the culm to facilitate wind pollination of the flowers. The usual type of inflorescence in grasses is the *panicle* (fig. 3), which consists of a main axis or *rachis* bearing, at intervals, one to several or sometimes many slender commonly scabrous (roughened) *branches*. Usually small but sometimes large flower-bearing structures called *spikelets* are borne toward the end of the branches. The slender stalks to which these spikelets are attached are called *pedicels*.

Panicles in grasses vary from open, sometimes large, airy *diffuse* types (fig. 76) to narrow or *contracted* forms (fig. 38). Dense, thick, cylindrical or nearly cylindrical panicles are referred to as *spikelike* (fig. 49). Some grasses have *compound* panicles as in Bermudagrass (*Cynodon dactylon*) or Large Crabgrass (*Digitaria sanguinalis*), where several spikes or spikelike racemes are radially arranged about the summit of the culm but still compose a branched inflorescence.

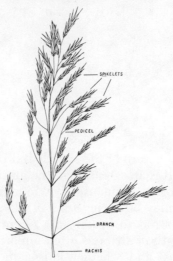

Fig. 3. The grass panicle

Racemes represent a reduction from the basic panicle structure and constitute a distinct inflorescence type. The spikelets are attached directly to the main axis by their pedicels without any intervening branches (fig. 4). The raceme in its simplest form, that is, solitary on the culm, occurs rarely. Two or more racemes compounded into a single terminal inflorescence is far more common (figs. 79, 80, 84, 91).

Fig. 4. The grass raceme (*Digitaria ischaemum*)

By further reduction, spikelets may be attached directly to the main axis by their glumes (see below). The resulting flowering structure is called a *spike* (fig. 5). Spikes may be solitary on the culm as in tribe Triticeae or several spikes may be compounded into a single inflorescence as in the tribe Chlorideae.

Fig. 5. The grass spike (*Lolium temulentum*)

SPIKELETS

The grass *spikelet* is the basic unit of the inflorescence and must be carefully studied in order to identify any given grass species. The arrangement of spikelets and the type of inflorescence formed, the number and type of florets they contain, and the way in which they fall away all are extremely important in grass classification. The very nature of the grass spikelet can reveal relationships between species and even genera.

An ordinary grass spikelet of 5 or more florets, as in the genera *Bromus*, *Festuca*, *Eragrostis* and *Leptochloa* (fig. 6), shows a strict 2-ranked (*distichous*) arrangement of these florets on a small jointed axis known as the *rachilla.* At maturity spikelets such as these fracture at the joints of the rachilla allowing the florets to fall separately with each carrying with it an internode of the rachilla (fig. 7).

At the base of the spikelet are two empty scales or bracts called *glumes* (fig. 6) In most grasses both glumes are present; the lower one is usually smaller, although both may be of equal length. Rarely one glume is suppressed or both glumes are completely absent. In many grass spikelets the glumes are persistent on the

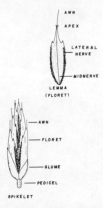

Fig. 6. The grass spikelet (*Leptochloa fascicularis*)

pedicel or, in the case of spikes, on the rachis. The florets then fall away above the glumes at the breakup of the rachilla; if there is only one grain-bearing floret, then this falls away from its point of attachment just above the glumes. Sometimes, as in the tribe Paniceae, the whole spikelet falls away as a unit from the pedicel that bears it. Disarticulation (disjointing) then, occurs below the glumes. Glumes are distinctive by their length, width, texture, number of nerves, and presence or absence of awns, hairs and barbs.

Above the glumes are the bracted flowers called *florets* (fig. 6) which may be numerous, several, or reduced to one. Their number is important in making identifications on the generic and tribal level. Reduction in the number or size of florets, sterility, or some modification of the bract structure commonly are expressed in many grass spikelets. Reduced floret size, their shape, character, and position in relation to grain-producing ones are extremely valuable in classifying grasses. They will be discussed later as a preface to the key genera.

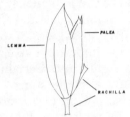

Fig. 7. The grass floret

Ordinary grain-producing florets (fig. 7) consist of two opposite bracts. The outer, larger and tougher bract is known as the *lemma* while the inner, rather delicate, commonly 2-nerved, usually shorter bract is called the *palea*. The *lemma* is the more distinctive and useful in grass determinations. It is variably nerved, provided with hairs or barbs, or it may be entirely smooth; whatever the condition the morphology of the lemma is consistent for a given species. In many grasses a bristlelike structure, the *awn*, (fig. 6), develops as a continuation of the midnerve and extends along or beyond the body of the lemma. The awn may arise anywhere along the back of the lemma or it may be terminal. The character of the awn and its presence or absence are an important diagnostic feature of grass spikelets. The nature of the palea is sometimes important in determining grass species, particularly when it is greatly reduced or absent as in the genus *Agrostis*, or when the hairs along the two nerves of the palea are stiff and comblike in appearance (pectinate) as in the genera *Bromus* and *Brachypodium*.

The lemma and palea enclose the *grass flower* (fig.

8). The flower consists of two minute, thin or fleshy scales called *lodicules* at the base, usually three *anthers*, each borne on long slender *filaments* (both constituting the *stamens*), and a central saclike *ovary* tipped by two feathery *stigmas*. The character of the lodicules varies from thin and nerveless to thick and nerved (or sometimes hairy) and serves to distinguish certain grass tribes. The function of the lodicules is to facilitate separation of the lemma and palea and thus to allow exposure of the anthers and stigmas to the wind. Stamen number is sometimes important in classification; any variance from 3 is significant. Certain annual species of *Festuca* have only a single stamen; in some other grasses there are two and in *Oryza* and the bamboos there are six stamens with each flower. Stigmas may be sessile on the ovary, elevated on a common style or the styles separate. The feathery stigmas very effectively trap pollen grains being carried by the wind. Upon fertilization the ovary matures into a 1-seeded fruit known as a *caryopsis* or, more commonly, *grain*. The ripened grain ordinarily remains permanently enclosed

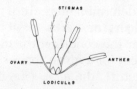

Fig. 8. The grass flower

between the lemma and palea and this whole structure, in actuality the floret, becomes the "seed" of grasses. The actual seed containing the embryo is enclosed in the grain.

The shape, size and character of the grain are of some value in grass determinations. Grasses of the tribes Triticeae and Bromeae have more or less oblong grains, tipped by a "brush" of hairs, and the seed contains simple starch grains. Grains may be loosely enclosed between the lemma and palea in some grasses as *Sporobolus, Eragrostis, Crypsis* and *Schismus*, adherent to the palea as in *Bromus*, or quite tightly enclosed in the tribe Paniceae. Some grasses as *Schismus* and certain *Eragrostis* species produce minute, transparent grains. Others, as Blasdale Bentgrass (*Agrostis blasdalei*), a rare native along the north coast of California, develop a liquid endosperm in the seed. The usual condition in grasses is production of a solid and starchy endosperm. Embryo size in relation to the endosperm separates grasses on the tribal level. For example, the Festuceae have relatively small embryos with large endosperm while in the Paniceae the embryo is much larger. The *hilum* or scar at the point of attachment of the ovary may be small and dotlike as in the grass tribe Festuceae while seeds of other tribes may have a narrow, elongate scar. Grooves or markings or lack of them may occassionally be of some importance in species determination. For example, some species of *Eragrostis* have a shallow groove along one side of the grain (fig. 62a).

EPIDERMAL STRUCTURES

The surfaces of leaves, stems, and parts of the inflorescence are often covered with small, hard, somewhat triangular-shaped *barbs*. A surface with barbs is rough to the touch and the condition is called *scabrous*. Some structures such as panicle branches, the upper surface and margins of leaves or the backs of lemmas may be strongly scabrous, that is, densely beset with barbs. Minute and perhaps widely spaced barbs produce a condition known as *scaberulous*, i.e., a barely noticeable roughness. Barbs are ordinarily pointed in the direction of the apex of the structure they cover and hence are *antrorse*. Backwardly or downwardly directed (*retrorse*) barbs are the exception and therefore significant in species determination.

The length and character of *hairs* on any structure is frequently of value in grass species determinations. The presence of any kind of hair is a condition of *pu-*

bescence. Minute hairs requiring some magnification to see is a *puberulent* condition. Most hairs in grasses are soft and without swollen bases. Stiff or stout (*hispid*) hairs sometimes have swollen bases as in the genera, *Digitaria, Panicum*, and *Echinochloa*. In a *ciliate* condition hairs occur along the margin of any structure. When these hairs are stiff it is further qualified by the term *pectinate* (comblike), that is, pectinate-ciliate The hairs so far discussed are the *macrohairs* which we easily see with the naked eye or some little magnification. On the leaf epidermis of many grasses are *microhairs* requiring use of some compound microscope to see. They are usually 2-celled but with a distinctive shape for various genera or tribes of grasses. In the subfamily Festucoideae microhairs are absent.

The absence of any kind of macrohair is a *glabrous* condition. *Smooth* refers to the absence of both barbs and hairs.

Glands are seldom present in grasses but when developed are significant. They are modified cells of the epidermis which apparently excrete a watery substance that eventually becomes sticky. Stinkgrass (*Eragrostis cilianensis*) is well supplied with glands, the excretions of which impart a peculiar odor to the plant. In the extremely rare grass genera *Orcuttia* and *Neostapfia* glandular excretion is common. The whole plant, in species of each genus, becomes more or less sticky as it matures and possesses a distinctive odor. The function of glands is not well understood.

SEED DISPERSAL

The primary means of seed dispersal in grasses is by gravity, the florets simply falling to the ground about the parent plant. Most florets are more or less buoyant and may be carried a short distance by water. In the genus *Paspalum* the spikelets are essentially boat-shaped, i.e., rounded on one side and flat on the other, thus well adapted for water transport. In some species of the genus, long silky hairs occur along the margin of the spikelet ensuring even greater buoyancy and perhaps longer transport in water.

Spikelets or florets with light grains and surrounded by long silky hairs are adapted for wind transport. Many species of the genus *Andropogon* disperse their fruits in this manner. A further adaptation for wind transport often lies in the character of the inflorescence. In certain grass species such as *Agrostis avenacea, Panicum capillare,* and *Panicum hillmanii* the panicle is broad, diffuse and airy. At maturity the whole panicle breaks away by a fracturing of the culm below, and is carried by wind along the surface of the ground. This tumbleweed behavior effectively scatters spikelets along the way. The long awns and glumes of the bushy squirreltail spike (fig. 26) allow that spike to be carried along by wind as a small tumbleweed. The spike finally breaks up and the florets fall away as it rolls over the ground. In strong winds the spikes, or at least portions of them, may become airborne for considerable distances.

Long awns, sharp-pointed florets, and barbed or hairy lemmas are remarkable adaptations for transport of seeds in animal hair. The sharp-pointed character allows easy penetration into the hair and the usually upwardly pointed barbs on awns and lemmas prevent easy withdrawal. Seeds then may be carried in hair or in man's clothing or equipment for long distances before finally being dislodged. The rare development of spines in grasses occurs in the fruits of *Cenchrus* (figs. 89, 90). These sharp-pointed spines are backwardly barbed so that upon penetration of skin (very painful), animal hair, or man's clothing, they are difficult to remove.

Man himself is instrumental in dispersing grass seed over the entire world by planting his grass crops, the cereals, the animal forages and the turfgrasses. The cereals depend solely upon man for maintenance since they soon succumb if left untended in competition with natural resident vegetation. This may also be true of the many animal forages in "tame" pastures. In such species there has been much selection for yield and disease resistance and general high quality as animal feed. Some, however, do escape from pastures and successfully compete with native plants.

USES OF GRASSES

Grasses were a fully developed, diversified and an established group of plants before the advent of man upon the earth. No comparable group of flowering plants have been so useful in his existence. As man explored and populated the tropical and temperate regions of the world he discovered and improved on the staple food grains of sorghum, a variety of millets, rice, wheat, barley, rye, oats and corn. Sugar cane provided a quick source of energy, raw sugar.

The domestication of herbivorous animals as a source of food and labor of necessity involved intensive use of natural grasslands as well as eventual development of the high protein concentrated "tame" pastures. Grass as forage for cattle, sheep, goats, horses, and other livestock is unexcelled. The high level of protein and adequate supply of minerals and vitamins in grasses ensure proper nutrition for the grazing animal and a rich source of meat and dairy products for man.

All of the cereals are grown in California, most of them in large acreages. Millions of acres of natural range provide much grazing land for domestic animals as do hundreds of thousands of acres of irrigated pastures. In the diagnoses of genera and species to follow, all of the important forage grasses in both areas are described.

Certain sod-forming grasses are useful as turf. In the northern half of California select species of the genera of *Poa*, *Agrostis*, and *Festuca* are best adapted as turfgrasses. Additionally, in the southern half of the

State, the warm-season grasses such as Bermudagrass (*Cynodon dactylon*) and Kikuyugrass (*Pennisetum clandestinum*) are sometimes quite useful as ground cover. There are hundreds of thousands of acres of turfgrass in California devoted to landscape ground cover primarily as lawns. Golf courses and athletic fields account for much additional acreage. Those species useful as turf are included in the diagnoses of genera and species to follow.

As subjects of ornamental plantings, exclusive of lawns, certain select grasses are used because of the special effects created. Bamboos, especially the hardy *Phyllostachys* species (pl. 6a) are widely used. Pampasgrass (*Cortaderia selloana*) with its striking mound of foliage and beautiful plumelike panicles (pl. 6f) is useful in certain landscape plantings. The attractive panicles of *Pennisetum* species (pl. 8a, b) and actually a wide variety of grasses often lend distinctive character or provide variety in ornamental plantings. A useful strain of *Festuca ovina*, known as Blue Fescue, is attractive as a ground cover because of the unique tufts of blue-green foliage.

Grass inflorescences are decorative in dried flower arrangements. There are many distinctive forms and considerable variety is possible in the dried bouquets. Dried cereal heads (spikes and panicles) are especially graceful but there is also a great variety to be found in our native California grass species; unfortunately many of them shatter easily unless they are collected early.

Grains of certain species of *Bromus* and quite likely many grasses in the tribe Triticeae provided food for the native Indian populations of the state. Certain stout and tough grasses such as Deergrass (*Muhlenbergia rigens*) may have been used by the Indians in making baskets, as were the slender but durable stems of Common Reed (*Phragmites australis*). The Reed was used for other types of utensils, and perhaps for implements of various kinds, including arrow shafts.

IDENTIFICATION OF GRASSES

Grasses constitute a natural plant group or family. There is remarkable similarity of characteristics as to growth habit, nature of stems and foliage, and floral structure spread over many thousands of species. Identification of species or any category depends upon a series of correlated characters. Differences exist in all areas of the grass plant except in the uniform fibrous roots. To determine any given species of grass, a series of opposing or contradictory characteristics are presented as a plant *key*. Through a process of elimination a combination of characters is chosen as best applying to the specimen to be identified and the plant is thereby "keyed out". Keys may be provided for families, tribes, genera, species or even varieties. In this guide there are keys to genera and species.

Every known plant (and animal) species bears a scientific name that indicates its identity and its relationship to similar kinds. That name is of two parts, the *genus* and the *species* (binomial nomenclature). The first letter of the generic name is always capitalized while that of the species is not (e.g., *Poa annua*). Both parts of the name are in Latin or else have been Latinized from another language such as Greek. The naming is controlled through international rules of nomenclature so such names are perfectly understandable to scientists everywhere in the world. Furthermore, the scientific name implies relationship. Species in any grass genera such as *Poa, Agrostis, Festuca, Panicum*, etc. all share common characteristics and are closely related.

RELATIONSHIPS IN GRASSES

The size and extent of the grass family is not precisely known and any figures used are necessarily estimates. Probably there are over 600 genera and over 10,000 species of grasses throughout the world. More grasses are being discovered as the botany of each region is more thoroughly investigated. As genera are carefully studied by monographs or revisions various species emerge or are combined.

In California there are 98 genera and 478 species of grasses. Many species and sometimes certain genera are of limited distribution or occur as scattered plants over a wide area. There are 303 native species and 175 species that have been introduced. Of these 322 species are perennial and 156 are annual.

There is, from the standpoint of practicality, a major reliance upon the external appearance and structure (external morphology) to readily identify any family, genus, species or variety of plants. Comparisons are made and relationships established by those characters one can readily see and assess. From the classical viewpoint, the inflorescence and its flowers is probably the most "relatable" mechanism because it is less subject to variable modifications imposed through sometimes rapid environmental changes. Many studies, besides classical external morphology, are bringing relationships of plants and plant groups into clearer focus. These studies are the major part of the science of *taxonomy*.

Our understanding of grasses and their interrelationships has markedly increased during the last forty years. Much taxonomic work has been done through use of the microscope. Details of the anatomy of grass stems, leaves, roots, embryos, and epidermis have become known and correlated with still other characteristics of the plant. Cytological investigations reveal certain basic chromosome numbers and sizes along with their organization into multiple sets (polyploidy).

There is much work yet to do among grasses relative to their genetics, though the crop cereals and sugar cane have been investigated thoroughly as to inheritance, transmission of characteristics, hybridization and reproduction. Studying of the distribution of protein fractions through electrophoresis is of considerable recent interest. The distribution of these fractions for each species or group of species follows a definite pattern and is of value in confirming or establishing relationships.

Biochemical studies have shown similarities among grasses in their germination responses to low oxygen levels and treatments with chemicals (especially with isopropyl-n-phenyl carbamate). In the grass tribes Festuceae and Eragrosteae the nature of serological responses suggest close relationships between the species of genera in each tribe.

The distribution of grasses over the earth's land surface provides significant clues to grass relationships. The temperate genera prefer a cool growing season and tropical genera a warm or even a hot growing season. In each region the major genera have evolved a series of species in such a climatic condition with little interchange of germ plasm except along the overlap areas. The two climatic zones roughly form the basis of two major divisions of the grass family—the festucoid line (resembling *Festuca* as typical) primarily in the temperate regions and the panicoid (resembling *Panicum* as typical) line primarily of tropical regions. Certain grasses from either zone have become adapted to conditions of the other. California has the best representation of temperate genera but also has some genera of tropical origin.

The evidences and correlations from these many lines of research makes possible a more or less natural alignment of genera sharing common characteristics. A group of such related genera is known as a *tribe*. The tribal name is derived from the typical genus of that group, as an example, the tribe Festuceae from the genus *Festuca*.

The genera of grasses treated in this book are numbered and placed in the following tribes. In the descriptions of genera and species this numerical order is followed. In addition to listing genera by tribes, the tribes are further organized into larger categories, the *subfamilies*. This arrangement is based upon the evaluation of many characters. It is presented here as an expression of the probable relationship of the genera and tribes common in California.

Subfamily FESTUCOIDEAE

TRIBES
Festuceae
— *Festuca* (1), *Lolium* (2), *Poa* (3), *Dactylis* (4), *Briza* (5), *Lamarckia* (6), *Cynosurus* (7).

Meliceae
— *Melica* (8), *Glyceria* (9).

Bromeae
— *Bromus* (10), *Brachypodium* (11).

Triticeae
— *Triticum* (12), *Aegilops* (13), *Hordeum* (14), *Sitanion* (15), *Elymus* (16), *Agropyron* (17), *Secale* (18).

Aveneae
— *Avena* (19), *Aira* (20), *Holcus* (21), *Deschampsia* (22), *Koeleria* (23), *Trisetum* (24), *Anthoxanthum* (25), *Agrostis* (26), *Calamagrostis* (27), *Polypogon* (28), *Gastridium* (29), *Phleum* (30).

Phalarideae
— *Phalaris* (31).

Stipeae
— *Stipa* (32), *Oryzopsis* (33).

Ampelodesmeae
— *Ampelodesmos* (34).

Subfamily BAMBUSOIDEAE

TRIBES
Bambuseae
— *Phyllostachys* (35).

Subfamily ORYZOIDEAE

TRIBES
Oryzeae
— *Oryza* (36).

Ehrharteae
— *Ehrharta* (37).

Subfamily ARUNDINOIDEAE

TRIBES
Arundineae
— *Arundo* (38), *Phragmites* (39), *Cortaderia* (40).

Danthonieae
— *Danthonia* (41), *Schismus* (42).

Aristideae
— *Aristida* (43).

Subfamily ERAGROSTOIDEAE

TRIBES
Eragrosteae
— *Eragrostis* (44), *Sporobolus* (45), *Crypsis* (46), *Muhlenbergia* (47).

Chlorideae
— *Chloris* (48), *Cynodon* (49), *Leptochloa* (50), *Hilaria* (51),

Aeluropodeae
— *Distichlis* (52)

Orcuttieae
— *Neostapfia* (53).

Subfamily PANICOIDEAE

TRIBES
Paniceae
— *Panicum* (54), *Digitaria* (55), *Eriochloa* (56), *Echinochloa* (57), *Paspalum* (58), *Pennisetum* (59), *Setaria* (60), *Cenchrus* (61).

Andropogoneae
— *Andropogon* (62), *Miscanthus* (63), *Sorghum* (64), *Zea* (65).

DISTRIBUTION OF GRASSES IN CALIFORNIA

Over half of the grass species growing in California occur at low elevations in a Mediterranean climate. This is the VALLEY-FOOTHILL area and comprises the grassland community in California. In the mountains, the MONTANE area, and along the coast, the COASTAL area, grasses compete with the forest and the brushland and consequently are best developed in the open places. The DESERT area has a relatively low density of grasses owing to the low rainfall and perhaps to overgrazing by domestic animals in the spring period.

The distribution of native species depends upon the climate, the nature of the soil, special character of the habitat and type of plant association. The naturalized grasses are commonly widespread since they are seldom influenced by all of the factors limiting the spread of the native species.

THE VALLEY-FOOTHILL AREA

This area is typified by the Central Valley and the surrounding foothills but further extends toward the coast into southern California. This corresponds to the *California grassland* and supports over 65 percent of the livestock grazing in the state. The principal forage plants are annual grasses and other annual herbaceous plants. The character of this grassland has changed from its pristine condition.

Before 1769 only native deer, elk, and antelope grazed over a grassland dominated by perennial species and best developed in the northern half of California. On the drier soils of the valley plains and foothill slopes were found: Purple Stipa (*Stipa pulchra*), Nodding Stipa (*Stipa cernua*) (pl. 1h), Pine Bluegrass (*Poa scabrella*), Blue Wildrye (*Elymus glaucus*), California Melic (*Melica californica*), Small-flowered Melic (*Melica imperfecta*), California Brome (*Bromus carinatus*), Junegrass (*Koeleria cristata*), and California Oatgrass (*Danthonia californica*). Big Squirreltail (*Sitanion jubatum*) was fairly common on gravelly flats, slopes or ridges and generally infertile soils where other grasses would be scarce.

The rich loams at the edge of the tule marshes along the Sacramento and San Joaquin rivers supported the following perennials: Creeping Wildrye (*Elymus triticoides*), Slender Wheatgrass (*Agropyron trachycaulum*) Bearded Wheatgrass (*Agropyron subsecundum*), Meadow Barley (*Hordeum brachyantherum*), Deergrass (*Muhlenbergia rigens*), Pinegrass (*Calamagrostis rubescens*), and Prairie Wedgegrass (*Sphenopholis obtusata*). In the marshes or at the edge of the rivers were these perennials: Tufted Hairgrass (*Deschampsia caespitosa*), Spike Bentgrass (*Agrostis exarata*) Reed Canarygrass (*Phalaris arundinacea*), Knotgrass (*Paspalum distichum*), Foxtail Barley (*Hordeum jubatum*), Common Reed (*Phragmites australis*), Rice Cutgrass (*Leersia oryzoides*), and an annual, American Sloughgrass (*Beckmannia syzigachne*).

On the alkaline flats of the Central Valley these perennials flourished: Saltgrass (*Distichlis spicata*), Alkali Sacaton (*Sporobolus airoides*), Nuttall Alkaligrass (*Puccinellia nuttalliana*) and in moist areas, Scratchgrass (*Muhlenbergia asperifolia*). Some annuals also grew on alkali namely, California Alkaligrass (*Puccinellia simplex*), Alkali Barley (*Hordeum depressum*) and in moist areas, Bearded Sprangletop (*Leptochloa fascicularis*).

Rainwater collecting basins (vernal pools) or "hog wallows" were a common feature over most of the Central Valley flats and rolling plains. Many unusual plants came to be associated with, and restricted to, this habitat. The grasses in these pools were all annuals: Annual Hairgrass (*Deschampsia danthonioides*), Pacific Foxtail (*Alopecurus saccatus*), Lemmon Canarygrass (*Phalaris lemmonii*), Small-leaved Bentgrass (*Agrostis microphylla*), Sixweeks Bentgrass (*Agrostis exigua*),

Annual Semaphoregrass (*Pleuropogon californicus*) and, in the largest pools, the unusual grass genera, *Orcuttia* and *Neostapfia*.

Widely scattered remnant areas of the California grassland still exist in the Central Valley but possibly not for long. Some species such as Prairie Wedgegrass, Pinegrass, Junegrass, Small-flowered Melic and California Oatgrass have long since disappeared from the Central Valley but are found in the surrounding foothill regions. In Solano County good stands of Purple Stipa grow in the vicinity of Dozier Station (pl. 1g) and south of Denverton. Also, near Dozier Station, several important vernal pools contain all of the annual species of the pristine condition.

In the foothills surrounding the Central Valley and the hills along the coast in central and southern California most of the perennials of the California grassland are still present, though usually in sparse stand. Prairie Wedgegrass has nearly disappeared and California Oatgrass occurs mostly in the coastal counties.

Annual grasses were not generally abundant on the dry plains or hillslopes in the pristine grassland. Small Fescue (*Festuca microstachys*), Sixweeks Fescue (*Festuca octoflora*) and Scribneria (*Scribneria bolanderi*) were best developed on infertile, stony soils and still occur there. In southern California and on the sandy soils of the San Joaquin Valley Arizona Brome (*Bromus arizonicus*) was an important annual species. Howell Bluegrass (*Poa howellii*) was, and still is, common on wooded slopes and at the edge of the brushlands. Broad-leaved (forbs) annuals were probably abundant in the original grassland. These annuals were from such families as the Sunflower (Compositae), Legume (Leguminosae) Borage (Boraginaceae), Figwort (Scrophulariaceae), Poppy (Papaveraceae) and many others.

The California grassland has long been exposed to the slow climatic changes of diminishing rainfall and increase of temperature with the eventuality of a hot, dry summer. The character of the grassland was dramatically altered when European livestock entered southern California in 1769 with the Spanish soldiers and the missionary Fathers. Cereals and fruits were soon imported and grown around the missions established along the coast. Other plants were also introduced—some purposely and some accidentally. The accidents were the casual weeds which were transported in animal hair, packing materials, ships' ballast or in soil surrounding fruit cuttings. Most of these weeds were annuals, and many were grasses: Red Brome (*Bromus rubens*), Downy Chess (*Bromus tectorum*), False Foxtail Fescue (*Festuca myuros*), European Foxtail (*Festuca bromoides*), Foxtail Fescue (*Festuca megalura*), Hare Barley (*Hordeum leporinum*), Glaucous Barley (*Hordeum glaucum*) Nitgrass (*Gastridium ventricosum*), Purple Falsebrome (*Brachypodium distachyon*) and Silver Hairgrass (*Aira caryophyllea*). The early Spaniards may have directly imported seeds of these annuals: Wild Oats (*Avena fatua*), Slender Wild Oats (*Avena barbata*), Annual Ryegrass (*Lolium multiflorum*) and perhaps even Soft Chess (*Bromus mollis*), and Ripgut (*Bromus diandrus*), all of these species having proven their worth as animal forage elsewhere. The important annual legume, Bur Clover (*Medicago polymorpha*) and the Filarees (*Erodium* species: *E. botrys*, *E. obtusiplicatum*, *E. cicutarium*, and *E. moschatum*) were probably directly imported as proven and valuable sheep forage. Without a doubt, impurities, which were weedy species of far less forage value, were carried in the imported seed.

The forage and weedy annuals soon became well established in the mission areas and eventually spread inland as the animal grazing area advanced. Great herds of animals grazed around the missions and still greater numbers on the large ranchos. Many areas were continually overgrazed, which placed severe stress on the native vegetation. The aggressive, well-adapted European annuals fared well—so well in this Mediterranean climate that they eventually became dominant.

In the Central Valley, long grazed by the Spanish herds, cereal farming in the mid 1800s quickly removed much native grassland. Later more diversified and extensive agriculture removed still more native grassland. Improvement of the natural dry pastures, in the last 30 years, has introduced several high-yielding annual Clovers (*Trifolium* species, *T. hirtum*, *T. subterraneum* and *T. incarnatum*) along with a perennial, Hardinggrass (*Phalaris tuberosa* var. *stenoptera*) both types of plants from the Mediterranean region. The current management practices of the valley and foothill ranges favors development of the best resident annual forage grasses or recommends seeding superior forage species.

THE MONTANE AREA

The montane area comprises that portion of California essentially above 3000–4000 feet, exclusive of the deserts which lie east and southeast of the high mountain chains in the state. The dominant plant cover is coniferous forest or, in some areas, mixed forest, either type usually with an undercover of shrubs of various kinds.

Since grasses are best developed in open areas, the meadows, flats, valleys and rocky slopes become their major habitats. Fairly open stands of conifers on the mountain slopes facing the desert area to the east support good stands of grass species. The summer grazing areas in the mountains is where the grass is most abundant. The cool summer and fall provide a long period of lush and succulent forage. Most of the montane grasses are native perennials though a number of European perennial species are well adapted here: Timothy (*Phleum pratense*), Orchardgrass (*Dactylis glomerata*), Redtop (*Agrostis alba*), Smooth Brome (*Bromus inermis*) Kentucky Bluegrass (*Poa pratensis*), Tall Fescue (*Festuca arundinacea*) Tall Oatgrass (*Arrhenatherum elatius*), and several Wheatgrass (*Agropyron*) species.

The important mountain genera have long been associated with cool climates and moist soils during the growing season. They are residual in habitats that developed at the margins of glacial ice during the recent Ice Age. The genera *Bromus, Melica, Koeleria, Stipa, Agropyron, Danthonia* and *Sitanion* occupied drier moraine-type soils and later spread over exposed dry slopes. *Agrostis, Calamagrostis, Poa, Deschampsia, Trisetum, Glyceria, Alopecurus,* and *Muhlenbergia* grew on moist, meadow-type soils and still occupy those areas.

All of the montane genera, except possibly *Sitanion*, are valuable as livestock forage. Annual grasses, in numbers of species, are relatively few in the mountains. Occasionally a species that has become naturalized, as Downy Chess (*Bromus tectorum*), is now widespread over more or less dry soils through much of the Sierra Nevada.

Undoubtedly the density of many of the perennial species has been affected by livestock grazing year after year. Weak Mannagrass (*Glyceria pauciflora*), for example, has literally been grazed out of many areas, having enjoyed far greater abundance before livestock were introduced into mountain meadows. On drier soils, particularly on the east slope of the Sierra and Cascades, species of the *Festuca, Agropyron* and *Stipa* genera were probably in less density during the periods of more rainfall but became more abundant with increasing aridity and reached some stability just before the introduction of domestic animals. Under heavy livestock grazing some areas were denuded of preferred species and rapidly declined with an increase of Squirreltail (*Sitanion hystrix*) and Downy Chess (*Bromus tectorum*), two less desirable species. Now careful management of bunchgrass rangelands is the rule with consideration given to stocking rates, range condition, rotational grazing, and even recreational activities.

THE COASTAL AREA

The coastal climate of California has relatively uniform temperatures cool winters and cool summers in the north to warmer in the south. Rain occurs from the fall through spring but the summers are rainless. Some soil moisture is derived from the summer fogs. The amount of rainfall varies from over 80 inches in Del Norte County to as little as 8 inches in San Diego County. The vegetation also varies from the dense redwood forests in the north to desert scrub in the south.

The north coast of California is rich in perennial species of grasses (pl. 2). The summer coolness favors development of several genera that are common at medium to high elevations in the interior mountains namely *Agrostis, Deschampsia, Calamagrostis* and *Danthonia*. There are several endemic species of *Agrostis* along the coast, growing nowhere else in the world. They are California Bentgrass (*Agrostis californica*), Cliff Bentgrass (*Agrostis clivicola*), Blasdale Bentgrass (*Agrostis blasdalei*) and Point Reyes Bentgrass (*Agrostis aristiglumis*). All of them occur mostly on the bluffs facing the sea.

The common elements of the coastal slopes, meadows, and bogs are primarily introductions from Europe: the perennial Velvetgrass (*Holcus lanatus*), Sweet Vernalgrass (*Anthoxanthum odoratum*), Colonial Bentgrass (*Agrostis tenuis*) and Tall Fescue (*Festuca arundinacea*) plus a variety of naturalized annual grass species of the genera *Bromus, Lolium, Avena, Aira,* and *Briza*. Hairy Oatgrass (*Danthonia pilosa*), a weedy perennial from Australia, is now rather widespread along the coast north of San Francisco.

The open slopes and flats along the north coast are

prime grazing areas for dairy animals, cattle, and
sheep. The forested areas are poor in grass species.
There are, however, two unusual grasses, as their common name suggests, which grow in the dense shade of
the redwood forest. They are California Sweetgrass
(*Hierchloe occidentalis*) and California Bottlebrushgrass (*Hystrix californica*). As the forest disappears
along the coast, a coastal scrub, with Coastal Sage (*Artemisia californica*) as the dominant species, occupies
the coastal hills from around the San Francisco Bay
area south into Lower California. This shrub formation is of fairly low stature and somewhat open character. Though grasses do grow among the shrubs they
are seldom in dense stands.

Some of the native grasses common along the north
coast such as Pacific Reedgrass (*Calamagrostis nutkaensis*), Pacific Hairgrass (*Deschampsia holciformis*)
and California Bentgrass (*Agrostis californica*) eventually disappear in Santa Barbara County at about
Point Conception. On the other hand grasses predominantly of the southern coastal region such as Crested
Stipa (*Stipa coronata*), Giant Wildrye (*Elymus condensatus* and Small-flowered Melic (*Melica imperfecta*) extend as far north as Napa, Alameda and Lake
Counties respectively. The native Foothill Stipa (*Stipa lepida*) and Thingrass (*Agrostis diegoensis*) extend
along most of the coast as well as inland. Purple Stipa
(*Stipa pulchra*) and Nodding Stipa (*Stipa cernua*)
which occur inland in and about the Central Valley
reach the coast in Humboldt County and the Monterey
Bay area respectively and thence southward to San Diego County. The naturalized European annuals occur
in abundance along the entire coast.

THE DESERT AREAS

The desert areas of California lie east of the Cascade-Sierra Nevada crest, with the largest area comprising
the Mojave and Colorado deserts.

The dominant vegetation is of shrubs, and the grasses must compete with these plants for water. The grass stand is much sparser and is mostly native perennials. From the north two major species of grasses, Indian Ricegrass (*Oryzopsis hymenoides*) and Needle-and-Thread (*Stipa comata*) extend along the eastern California border south to Inyo County. Indian Ricegrass ranges further southward through the MoJave and Colorado deserts into northern Mexico. Downy Chess (*Bromus tectorum*), a European annual, occurs in the higher elevations of the Mojave Desert and is widespread through the Great Basin Desert ascending into the Sierra Nevada and Cascade mountains.

The most abundant and widespread native grasses in the Mojave and Colorado deserts are Big Galleta-grass (*Hilaria rigida*), Desert Stipa (*Stipa speciosa*) and Indian Ricegrass (*Oryzopsis hymenoides*). Several species of *Aristida* and *Bouteloua* occur in the Mojave Desert above 4000 feet in elevation near the Nevada border. Heavy spring grazing here over many years has practically removed the stands of Blue Grama (*Bouteloua gracilis*) and other choice grasses. Fluffgrass (*Tridens pulchellus*) is fairly common but is better developed beyond the California border. Saltgrass (*Distichlis spicata*) is common in the alkaline playas through all of the desert regions. Two eastern Mediterranean annuals, *Schismus arabicus* and *Schismus barbatus*, are now widespread in both the Mohave and Colorado deserts and adjacent lands. The unusual and uncommon, hairy, rhizomatous Desert Panicum (*Panicum urvilleanum*) occurs along the dry river beds or large washes in both deserts.

Grasses in the desert, whether in fairly good or scattered stands are of great value in holding soil in place. Any destruction or absence of grass greatly accelerates erosion. Reseeding projects are largely out of the question because of scarce and unpredictable rainfall. Indian Ricegrass (*Oryzopsis hymenoides*), the most important desert species, has exceptionally hard seeds of long viability to survive adverse or nil rainfall seasons.

CROPS AS AN AREA OF GRASS DISTRIBUTION

The vast acreages of croplands in California provide a habitat for certain weeds that might never survive under the undisturbed native conditions. This is particularly true of the warm-season grass genera, which receive their moisture in the summer period through irrigation of the crops. *Digitaria, Echinochloa, Setaria, Eragrostis, Leptochloa, Panicum, Chloris,* and *Cenchrus* are common summer weedy grass genera.

A KEY TO THE GRASS GENERA

In using the following key considerable attention must be given to the details of the grass spikelet and how those spikelets are arranged in the inflorescence. Of particular importance is the position and character of sterile florets, if present. As a general rule, in spikelets with 2 or more florets, those with a well-developed lemma and palea are fertile (bisexual) while smaller or greatly reduced ones are sterile (neuter or occasionally with stamens only). Any structures quite different from the grain-producing floret either above or below (except the glumes) are to be regarded as modified or sterile florets producing no grains. Sterile florets ordinarily fall away from (or with) the spikelet (tribe Paniceae) always in close association with the fertile one(s). In many cases the sterile floret is represented only by a lemma or a structure interpreted as a lemma. It can be quite unlike the glumes or quite similar to the glumes (tribe Paniceae). Sometimes several sterile florets are condensed into an oblong structure known as a *rudiment* (*Melica species*). The pedicel of the rudiment is called a *stipe*. A few, or perhaps only one of the florets in the upper portion of a many- or even several-flowered spikelet may never produce grains even though both are with lemma and palea. The only distinction in such florets is the smaller size, resembling in most details those below. Glumes, which sometimes resemble sterile florets, are always basal.

In the key the number preceding the genus indicates its order of description as outlined on pp. 49–162. The number following the genus is the page of its description. (Following the descriptions of the larger genera keys are provided for the species.)

1. Spikelets borne on long or short pedicels, on the branches of a loose to dense panicle, occasionally in a loose raceme, or if in a cylindrical spikelike panicle then the spikelets crowded all around the axis.................Group 1 (p. 40)
1a. Spikelets sessile or sometimes very short-pedicelled and

placed along opposite sides or along one side of the axes of spikes or racemes..............Group 2 (p. 45)

Group 1

2. Spikelets with 1 or 2 dissimilar sterile florets *below* a terminal grain-producing (fertile) one.
    3. Sterile florets scalelike, closely appressed to the much larger fertile one.
        4. Glumes well developed, longer than the fertile floret; panicle commonly spikelike or sometimes loosely contracted..............
        31. *Phalaris* (p. 111)
        4a. Glumes wanting; panicle open, drooping...
        36. *Oryza* (p. 120)
    3a. Sterile floret(s) well developed, as long as or longer than the fertile one.
        5. Sterile florets 2 and quite unlike the enclosing thin glumes.
6. Panicle dense, spikelike, sterile florets awned from the back
   25. *Anthoxanthum* (p. 103)
6a. Panicle loose, sterile florets awned or awn-pointed from the tip ....37. *Ehrharta* (p. 120)
        5a. Sterile floret 1, resembling the upper glume in color, size and texture.
7. Panicle open................54. *Panicum* (p. 143)
7a. Panicle spikelike with long bristles attached to the pedicels below the spikelets forming sort of an involucre.
        8. Bristles persistent, the spikelets readily falling away from the panicle......60. *Setaria* (p. 155)
        8a. Bristles falling away with the spikelets........
        59. *Pennisetum* (p. 153)
2a. Spikelets with similar (though usually smaller) or sometimes dissimilar sterile florets *above* the grain-producing (fertile) one(s), rarely similar neuter or staminate *below* (*Phragmites* and certain *Eragrostis*), but if so then several fertile florets above, or spikelets with no sterile florets at all.
        9. Spikelets with a single fertile floret, sometimes accompanied by dissimilar, sterile, rudimentary floret(s) produced on a stipe or rachilla (some *Melica* species and *Holcus*) or the rudiment wanting leaving only the stipe or rachilla (*Calamagrostis* and a few *Agrostis* species)................Section 1 (p. 41)
        9a. Spikelets with two or more fertile florets....
        Section 2 (p. 42)

Section 1

        10. Sterile floret produced on a stipe or rachilla.
11. Spikelets falling entire from the pedicels; foliage softly hairy .... 21. *Holcus* (p. 96)
11a. Spikelets with glumes persistent on the pedicel, the floret(s) falling away at maturity; foliage finely scabrous.......... 8. *Melica* (p. 64)
        10a. Sterile floret wanting (sometimes a rachilla remaining but more commonly only the single fertile floret present).
12. Panicle dense, spikelike, cylindrical or nearly so; lemmas relatively thin.
    13. Spikelets falling entire from their pedicels, glumes awned from the back; panicles soft and bristly.. 28. *Polypogon* (p. 108)
    13a. Spikelets with glumes persistent on their pedicels (if sometimes the glumes are tardily deciduous [in *Crypsis*] then these not awned).
        14. Glumes tough, longer than the enclosed floret; erect perennials or annual.
            15. Glumes truncate, the midrib prolonged into a stout stiff awn; perennial........ 30. *Phleum* (p. 110)
            15a. Glumes long-tapering, somewhat swollen at the base; annual .... 29. *Gastridium* (p. 109)
        14a. Glumes rather thin, shorter than the floret; low spreading annual.............. 41. *Crypsis* (p. 131)
12a. Panicle open to narrowly contracted but not dense and cylindrical; lemmas thin to tough and sometimes hardened (*Stipa, Oryzopsis* and *Aristida*) at maturity.
16. Glumes ordinarily shorter than the lemma.
    17. Grain loosely enclosed between the lemma and palea; lemma smooth and awnless....45. *Sporobolus* (p. 130)
    17a. Grain scarcely loosely enclosed; lemma hairy or scabrous, short-pointed to long-awned at the apex........ 47. *Muhlenbergia* (p. 133)
16a. Glumes as long as or longer than the lemma.
        18. Lemma awnless or awned from the back, the awn fine, hairlike; florets not tough, leathery, nor hardened at maturity.
            19. Lemma callus with conspicuous tufts of long hairs; palea well developed; rachilla a hairy bristle..........27. *Calamagrostis* (p. 107)

19a. Lemma callus hairs greatly reduced to almost wanting; palea poorly developed, usually as a thin scale at the base of the lemma (in *Agrostis avenacea* the palea well developed, the rachilla a hairy bristle................
26. *Agrostis* (p. 103)

18a. Lemma awned from the tip, the awn commonly stout; lemma tough and leathery, becoming hardened at maturity.

20. Awn divided into three divergent segments..........43. *Aristada* (p. 126)

20a. Awn simple.

21. Awn commonly curved, rarely straight, early falling away from the floret; callus short, obtuse....33. *Oryzopsis* (p. 117)

21a. Awn twisted below and once or twice bent above, ordinarily persistent on the floret; callus mostly sharp-pointed....
32. *Stipa* (p. 113)

Section 2

22. Culms semi-woody canes over 2 meters tall; leaves rather evenly distributed along culm with no basal tuft; blades 2-6 cm. broad.

23. Rachilla hairy, lemmas glabrous...............
39. *Phragmites* (p. 121)

23a. Rachilla glabrous, lemmas hairy...............
38. *Arundo* (p. 121)

22a. Culms not semi-woody; leaves distributed along culm or many of them basal; blades less than 2 cm. broad.

24. Lemmas fan-shaped, many-nerved, glandular; rachis prolonged above the panicle........
53. *Neostapfia* (p. 143)

24a. Plants not as above.

25. Low, mat-forming, gregarious perennial with scaly, yellowish rhizomes; leaves with rather stiff and distichous blades and closely overlapping sheaths; plants of alkaline or saline soils...............
52. *Distichlis* (p. 141)

25a. Plants not as above.

26. Robust clumps; blades narrow, elongate with saw-tooth margins; culms over two meters tall; panicle large, plume-like, ordinarily silvery-white......40. *Cortaderia* (p. 122)

26a. Plants not as above.

27. Glumes well developed, one or commonly both as long or longer than the lowermost floret, often one or both glumes as long as or longer than *all* of the florets; lemmas awned from the back or awnless....Part 1 (p. 43)

27a. Glumes shorter than the lowermost floret; lemmas
awned from the tip or awnless................Part 2

Part 1

28. Spikelets 5- or more-flowered; ligule a fringe of
hairs; collar with a tuft of long hairs on either side.
29. Low annuals of arid regions; lemmas awnless
or nearly so, 2-3 mm. long................
42. *Schismus* (p. 124)
29a. Perennial bunchgrasses; lemmas awned, the
awn flat and often loosely twisted below,
sharply bent to one side above, lemmas also
exceeding 3 mm. in length, the points on
either side of the notch commonly attenuate
as short awns......41. *Danthonia* (p. 122)
28a. Spikelets ordinarily 2–3-flowered, but if rarely 4–5-
flowered then the ligule a membrane; collar without a tuft of long hairs on either side.
30. Glumes 2 cm. or more long; lemmas 1.5
cm. or more long, conspicuously long-
awned, the awn stout and tightly twisted
in the lower part....19. *Avena* (p. 93)
30a. Glumes and lemmas not as above, awned
or awnless.
31. Lemmas keeled on the back, V-shaped in cross section, the
awn arising from above the middle or awnless.
32. Lemma awnless or at most with a very short straight
awn; rachilla glabrous or minutely hairy............
23. *Koeleria* (p. 100)
32a. Lemma usually conspicuously awned, but if awnless in
some species, then the rachilla prominently hairy....
24. *Trisetum* (p. 100)
31a. Lemmas rounded on the back, awned from below the
middle, the awn slender and hairlike.
33. Lemma blunt and irregularly notched at the apex.
22. *Deschampsia* (p. 97)
33a. Lemma acute with 2 slender bristlelike teeth at
the apex.....................20. *Aira* (p. 95)

Part 2

34. Robust perennial forming large clumps; culms
solid, 2–3 meters tall bearing large somewhat
1-sided panicles 20–50 cm. long..........
34. *Ampelodesmos* (p. 119)
34a. Plants not as above.

35. Spikelets of two different kinds, sterile and fertile intermixed in a dense panicle.
36. Panicle oblong, yellowish or purplish tinged; fertile spikelet surrounded by a cluster of sterile spikelets, the whole group falling away as a unit............6. *Lamarckia* (p. 62)
36a. Panicle ovoid, greenish to whitish, densely bristly; only the florets of the fertile spikelet falling away..............
7. *Cynosurus* (p. 63)
35a. Spikelets all alike in an open or contracted panicle or sometimes in a simple raceme.
37. Spikelets large, ordinarily more than 1.5 cm long, if smaller then the grain concavo-convex in cross section and adherent to the palea.
38. Glumes stiff, the lower and shortest one 5-nerved, the upper and longer one 7-nerved; palea as long as the lemma body, stiffly ciliate along the margins in the upper half....11. *Brachypodium* (p. 77)
38a. Glumes not especially stiff, the lower one 1–3-nerved, the upper 3–5-nerved; palea commonly shorter than the lemma body and though ciliate, not stiffly so...............10. *Bromus* (p. 67)
37a. Spikelets commonly less than 1.5 cm. long but if rarely this length then the grains are not concavo-convex in cross section nor adherent to the palea.
39. Spikelets densely crowded and nearly sessile in 1-sided clusters at the ends of stiff panicle branches............4. *Dactylis* (p. 60)
39a. Spikelets not as above.
40. Lemmas about as broad as long, horizontally disposed in the spikelet......
5. *Briza* (p. 61)
40a. Lemmas not as above.
41. Lemmas 3-nerved; glumes, lemma and grain falling separately from the spikelet, palea more or less persistent on the non-disjointing rachilla..........44. *Eragrostis* (p. 128)
41a. Lemmas 5–9-nerved; glumes persistent on the pedicel, the rachilla disjointing and the florets falling away at maturity.
42. Lemmas ordinarily awned (awnless or nearly so in some species), the florets somewhat divergent from each other at maturity exposing the rachilla; blades commonly tightly rolled..........1. *Festuca* (p. 49)
42a. Lemmas ordinarily awnless (awned in some *Melica* species), the florets rather closely overlapping and not exposing the rachilla; blades ordinarily flat.
43. Lemmas 5-nerved, the lateral nerves sometimes indistinct....................3. *Poa* (p. 57)

43a. Lemmas 7–9-nerved, the nerves all prominent.
    44. Lateral nerves of lemma convergent toward the apex; plants of dry habitats..........
        8. *Melica* (p. 64)
    44a. Lateral nerves of lemma parallel with the midrib; plants of wet habitats................
        9. *Glyceria* (p. 66)

Group 2

    45. Culms perennial, woody; leaves with a petiolelike constriction between blade and sheath, the sheaths eventually deciduous; spikelets sessile in the axils of bracts....35. *Phyllostachys* (p. 119)
    45a. Culms, leaves, and spikelets not as above.
46. Spike or raceme 1, terminal on the culm; spikelets alternating on opposite sides of the rachis............Section 3
46a. Spikes or racemes several to many on the culm, either radiating terminally, scattered along, or placed on branches (paniculate) in the upper part of the culm; spikelets placed along one side of the axis of each spike or raceme........
    Section 4 (p. 46)

Section 3

47. Spikelets placed edgewise to the rachis, the inner glume wanting except the terminal spikelet. 2. *Lolium* (p. 55)
47a. Spikelets placed flatwise to the rachis, both glumes present in all spikelets.
    48. Spikelets in 3's placed side-by-side at each node of the rachis, the lateral spikelets neuter or staminate, the central spikelet grain-producing.
        49. Rachis commonly disjointing at the nodes, the triad of spikelets falling away entire with the internode of the rachis attached below. Annuals or tufted perennials..............
            14. *Hordeum* (p. 78)
        49a. Rachis not disjointing at the nodes, wavy, the triad of spikelets readily falling away as a unit. Plants perennial with rhizomes or stolons, of semi-arid or arid regions..........
            51. *Hilaria* (p. 140)
    48a. Spikelets solitary or in pairs at each node of the rachis.
        50. Spikelets one at each rachis node; glumes (except in *Secale*,) rather broad.

51. Florets 2; midrib and margins of the lemma stiffly hairy; glumes awl-shaped; rachis more or less disjointing at maturity................................18. *Secale* (p. 93)
51a. Florets mostly more than 2, but if 2 then lemmas, glumes, and rachis not as above.
    52. Tufted or rhizomatous perennials...................
                            17. *Agropyron* (p. 89)
    52a. Annuals
        53. Spikelets more or less cylindric................
                          13. *Aegilops* (p. 78)
        53a. Spikelets flattened.........12. *Triticum* (p. 78)
        50a. Spikelets ordinarily 2 at each node of the rachis; in *Elymus* sometimes more than 2 spikelets per node (*Elymus cinereus*, *Elymus condensatus*), or sometimes only 1 (*Elymus triticoides*) and if so then the glumes very narrow and sharp-pointed.
        54. Spikes extremely long-bristly because of long awnlike glumes and long-awned florets; rachis readily disjointing at its nodes.............
                          15. *Sitanion* (p. 84)
        54a. Spikes not as above.....16. *Elymus* (p. 86)

Section 4

    55. Spikelets unisexual, dissimilar, the staminate racemes organized into a large terminal panicle, the *tassel*, the pistillate spikes organized into an 8- to many-rowed *ear* which occurs on the stem below and enclosed by several overlapping spathes (*husks*)......65. *Zea* (p. 162)
    55a. Spikelets ordinarily bisexual but if some are staminate or neuter then these borne in the same raceme or spike and not in separate inflorescences in different parts of the plant.
56. Glumes well developed, becoming hardened, usually enclosing the thin lemmas and paleas; spikelets in pairs, the ordinary arrangement being one of the pair grain-producing and sessile, the other usually staminate or neuter and on a pedicel or sometimes only the pedicel is present; rarely the fertile spikelet on a pedicel and the pedicellate spikelet fertile (*Miscanthus*).
    57. Racemes of usually more than 5 joints and provided with long white or tawny hairs.
        58. Racemes many, 10 cm. or more long, in a fan-

shaped arrangement at the summit of the culm. .
63. *Miscanthus* (p. 160)

58a. Racemes solitary or several on a common peduncle, much less than 10 cm. long; peduncles usually several, ordinarily scattered along the upper part of the stem and enclosed at the base by a leaflike bract (spathe)................
62. *Andropogon* (p. 159)

57a. Racemes of 1–5 joints, without long hairs though sometimes inconspicuously short-hairy...............
64. *Sorghum* (p. 160)

56a. Glumes membranous, not becoming hardened, as long as or shorter than the floret(s); spikelets all alike even if sometimes apparently in pairs with one pedicelled and the other subsessile.

59. Spikelets with a single grain-producing floret.
60. Spikelets, or sometimes clusters of them, falling away as a unit; grain-producing floret enclosed between an upper glume and a glumelike sterile lemma, the lower glume greatly reduced or wanting; racemes or spikes commonly scattered on the upper part of the culm.

61. Racemes modified into spiny burs, the burs readily deciduous from the culm............61. *Cenchrus* (p. 158)
61a. Racemes not spiny burs and not falling from the culm.

62. Slender racemes radiating from the tip of the culm, sometimes with a secondary whorl below or these somewhat scattered, but if the latter then the lemma margin hyaline and not inrolled, the lemmas usually dark-colored....................55. *Digitaria* (p. 147)

62a. Racemes scattered along the upper part of the culm; lemma margin inrolled and not hyaline, the lemmas usually light-colored.

63. Spikelets with a dark-colored cuplike callus......
56. *Eriochloa* (p. 148)

63a. Spikelets not as above.

64. Ligule wanting; spikelets hispid..........
57. *Echinochloa* (p. 149)

64a. Ligule present; spikelets not at all hispid...
58. *Paspalum* (p. 152)

60a. Spikelets with glumes persistent, the floret(s) falling away; grain-producing floret solitary with a naked rachilla in *Cynodon*, or in *Chloris* with a modified angular or club-shaped rudiment above and not at all glumelike; racemes or

aggregated at the summit of the culm.
65. Low, mat-forming, perennial with well-developed rhizomes and stolons; rachilla without a sterile floret at the tip......
49. *Cynodon* (p. 137)
65a. Erect annual (*Chloris virgata*) or perennials; rachilla bearing 1 to several reduced florets appearing as an angular or club-shaped rudiment...............
48. *Chloris* (p. 137)
59a. Spikelets of 2 to several grain-producing florets. (Considerable attention must be given here in not confusing this genus with members of the tribe Festuceae, which it superficially resembles. It differs in the 3-nerved lemmas.)..........50. *Leptochloa* (p. 139)

DESCRIPTIONS OF THE GENERA AND SPECIES

1. Genus *Festuca* Fescue

Most *Festuca* species are perennial, but the genus includes some annuals. Blades commonly narrow, often involute, sometimes flat; sheaths, especially the lower, becoming brown and shredding into fibers in age; ligule ordinarily a short, truncate membrane; panicles open or contracted; spikelets several-flowered, mostly less than 1.5 cm. long, excluding the awns; glumes narrow, persistent; florets ordinarily divergent at least not remaining closely appressed to each other, readily shattering at maturity; lemma firm, more or less rounded on the back, awned from the tip, the awn straight; 24 species in California.

The perennial Fescues are valuable range forage grasses; they are used in pastures and lawns, or as ornamentals especially Blue Fescue (*F. ovina* var. *glauca*). The native and introduced species grow best in the mountains or along the north coast where the summers are cool.

The annual species (sometimes treated as a separate genus *Vulpia*) are commonly few-leaved, early maturing, and develop sharp-pointed, narrow, scabrous, long-awned florets in the spikelets. They are common as weeds about cities and towns, along roads, on disturbed or thin and rocky soils, in croplands and pastures, on rubbish dumps, land fills, etc. The florets are a nuisance to animals, often working their way into the ears, nose and mouth, also into hair and, sometimes, even penetrating the skin. On rangeland the annual species are poor forage, and an abundance of them indicates poor range condition.

1. Perennials.
    2. Blades flat or loosely rolled, smooth and shiny on the lower surface when fresh; lemma awn poorly developed, less than 2 mm. long or sometimes mucronate to awnless........................1. *F. arundinacea*
    2a. Blades mostly tightly rolled but if flat not smooth and shiny on the lower surface; lemma awn well developed.
        3. Collar region more or less pubescent; plants forming coarse clumps.............2. *F. californica*
        3a. Collar region glabrous.
            4. Blades scabrous and commonly glaucous...
                3. *F. idahoensis*
            4a. Blades smooth or nearly so, green or glaucous.
                5. Awn longer than the body of the lemma.
                    4. *F. occidentalis*
                5a. Awn shorter than the body of the lemma
                    5. *F. rubra*

1a. Annuals.
   6. Branches in the lower portion of the panicle widely divergent to reflexed; spikelets closely appressed to divergent in the upper portion; plants variously hairy to glabrous......
   6. *F. microstachys*
6a. Branches and spikelets rather closely appressed to the main axis and forming a narrow, sometimes whiplike and curved inflorescence; plants essentially glabrous.
   7. Lower glume 1/2 to 2/3 the length of the upper one..
   7. *F. bromoides*
   7a. Lower glume less than 1/2 the length of the upper one.
      8. Lemma without hairs along the upper margin...
      8. *F. myuros*
      8a. Lemma with hairs along the upper margin......
      9. *F. megalura*

1. *Festuca arundinacea* (pl. 2b)
Reed Fescue; Tall Fescue

Tufted perennial commonly forming large clumps, occasionally some short rhizomes produced; culms 60 cm. to as much as 2 meters tall; blades flat to sometimes loosely rolled in drying, shiny on the under surface, at least when fresh, dull and strongly ribbed on the upper surface, the base of the blade minute auriculate, the small auricles sparsely hairy; panicles 12–40 cm. long, open or loosely contracted; spikelets 10–15 mm. long; lemmas awnless or with short awn less than 2 mm. long.

Introduced from Europe and Asia. Now widely naturalized throughout California. It grows best on moist soils. At lower elevations this Fescue is used in irrigated pastures. Along roadsides and ditches and in areas of poor drainage it is quite abundant. At higher elevations Reed Fescue occurs in or about meadows, along roads, in areas of disturbed soil, and on fill soils. Because of its strong root system it is valuable in checking soil erosion.

2. *Festuca californica* California Fescue

Usually robust perennial clumps; culms rather stoutish; 60–140 cm. tall; foliage quite scabrous, pubescent about the collar; blades flat or involute; panicles open, 10–30 cm. long, the paired branches at each node of the rachis long, spreading, eventually drooping, bearing spikelets towards their ends; spikelets 4–5-flowered, lemmas commonly scaberulous, long-tapering or sometimes short-awned at tip.

Native. North Coast Ranges south to about Santa Barbara
County; occasional in the Sierra Nevada, commonly associated
with brushlands or heavily wooded areas, usually on serpentine
soils. In certain open brush associations California Fescue is the
dominant bunchgrass, sometimes in such density as to almost
form a carpet in the intervals between the shrubs. As a forage
grass it is fair to good depending upon its maturity. The fresh
new growth is palatable to grazing animals but as the stems
elongate and the panicle emerges it loses its appeal. The tall
stems and drooping branches of the panicle are quite attractive.

3. *Festuca idahoensis* (pl. 2c) Idaho Fescue

A densely tufted perennial, usually glaucous but some forms
green; culms 30–100 cm. tall; blades mostly basal, scabrous and
somewhat stiff; panicles usually loosely contracted; spikelets 8–10
mm. long, 5–7-flowered; lemmas firm, about 7 mm. long, tipped
with an awn usually 2-4 mm. long.

Native. From the San Francisco Bay area in the Coast Ranges
and Lake Tahoe in the Sierra Nevada north to Siskiyou, Modoc,
and Lassen counties, thence north to Canada and east to Colorado.

Idaho Fescue grows best in the mountains, where it is important as a range grass because of its abundance and utilization by
livestock. In the open pine forest-sagebrush association in northeastern California it is valuable in determining the carrying capacity of any given range in that area. Considerable management is necessary to maintain good stands since overgrazing soon
causes a rapid decline in the density of the species. In the Coast
Range foothills stands are rather sparse but with ascending elevation and association with coniferous forests, Idaho Fescue becomes more important in the ground cover.

4. *Festuca occidentalis* Western Fescue

Tufted perennial; culms 40–80 cm. tall; blades slender, mostly
basal and commonly smooth; panicles loosely contracted to somewhat open, the branches solitary or in pairs; spikelets 6–10 mm.
long, on hairlike pedicels; lemmas 5–6 mm. long, rather thin, the
awn as long as or longer than the lemma body.

Native. Mostly in the North Coast Ranges, Sierra Nevada, in
open coniferous forest, sometimes in oak woodland or mixed
forest of the north coast counties. Whenever abundant, the grass
is good forage for livestock.

5. *Festuca rubra* (fig. 9) Red Fescue

Loosely tufted perennial, in some forms short-rhizomatous; culms 35–70 cm. tall; blades mostly basal, smooth or nearly so; panicles 5–18 cm. long, commonly loosely contracted; spikelets 5–10 mm. long, purplish to green or glaucous; lemmas 5–7 mm. long, the awn 2–4 mm. long.

Native, or perhaps some forms introduced. Of variable habitats from moist meadow soils to drier sites. Red Fescue is best adapted to the north coastal region or to the higher mountains, rang-

Fig. 9. *Festuca rubra*; a, habit; b, c, glumes d, florets

ing throughout much of California where the summers are cool and there is adequate moisture during the growing season. Livestock readily graze it though it seldom occurs in dense stands. Certain forms are fine-leaved and esteemed in lawns.

6. *Festuca microstachys* (fig. 10) Small Fescue

Annual; culms 10–50 cm. tall; foliage sparsely hairy to glabrous; panicles reduced to 2–several short, stiff lower branches with several sessile or subsessile spikelets, the upper portion of the inflorescence with no branches, rather the divergent spikelets borne directly upon the culm axis; spikelets 2–3-flowered, the florets hairy and with a longer awn; glumes stiff, unequal, the lower 1-nerved, the upper 3-nerved.

Fig. 10. *Festuca microstachys*; spikelet and habit

Native. This species is included here because it is the oldest named (1848) of some 7 other similar species which may or may not be distinct entities. The characters separating them are disposition of hairs on the spikelets and habit of the panicle. All are similar in habitat, however, usually occupying thin, stony or sterile soils where competition is at a minimum. Small Fescue and its allies are distributed throughout California mostly below 5000 feet elevation.

7. *Festuca bromoides* Brome Fescue

Annual; culms 10–40 cm. tall; panicles not at all whiplike, 2–12 cm. long; spikelets several-flowered; lower glume a little longer than 1/2 that of the upper; lemma 7–8 mm. long tipped by an awn 10–30 mm. long.

8. *Festuca myuros* Rattail Fescue

Annual; culms 20–40 cm. tall; panicles narrow, whiplike, commonly curved, the branches closely appressed to the main axis; spikelets 4–5-flowered; glumes very unequal, the lower one 1–1.5 mm. long, the upper about 3.5 times as long; lemma scabrous towards the apex, not at all pubescent.

9. *Festuca megalura* (fig. 11) Foxtail Fescue

Similar to *Festuca myuros* and probably not specifically distinct, differing only in the presence of hairs along the upper margins of the lemma. The glumes are about 0.5 mm. longer. Undoubt-

Fig. 11. *Festuca megalura*; habit and spikelet

edly this grass was introduced into California during the mission period. It was well established about the Santa Barbara area by the time Gambel collected it and later Nuttall described it in 1848.

2. Genus *Lolium* Ryegrass

Annuals or rarely short-lived perennials; foliage glabrous, blades flat, shiny below when fresh; spikelets flattened, sessile and placed edgewise alternately on opposite sides of the rachis, widely spaced, the resulting spike commonly elongate; florets several to numerous per spikelet, readily deciduous upon maturity; lemmas rounded on the back, awned or awnless; in all but the terminal spikelet one glume completely suppressed, the other well developed, tough, strongly 3–5-nerved.

1. Glume ordinarily shorter than the spikelet.
   2. Lemmas, or at least some of them, awned..........
                                              1. *L. multiflorum*
   2a. Lemmas all awnless.................2. *L. perenne*
1a. Glume as long as or longer than the spikelet............
                                               3. *L. temulentum*

1. *Lolium multiflorum* (pl. 2d) Annual Ryegrass; Italian Ryegrass

Annual; culms 25–100 cm. tall, reddish at the base; blades flat with rather well-developed clawlike auricles at the base; spikes 10–40 cm. long; spikelets 10–20-flowered; lemmas, or at least some of them, awned.

Introduced from Europe. Now naturalized throughout much of California, mostly at low elevations. Annual Ryegrass is useful in erosion control and is frequently sown in brushlands ravaged by fire, or in roadbanks and fills, or along roadsides. It is used to provide a quick but temporary lawn about buildings and as forage on the valley and foothill rangelands, though it prefers deep, rich soils for maximum development and persistence. Frequently this Ryegrass is sown in irrigated pastures where it produces much succulent forage.

2. *Lolium perenne* (fig. 12) Perennial Ryegrass; English Ryegrass

Short lived perennial; culms 25–50 cm. tall; blades without auricles or sometimes with rudimentary auricles as short points;

Fig. 12. *Lolium perenne*

spikelets 6–10-flowered; lemmas awnless.

Introduced from Europe. Now naturalized throughout much

of California at low to medium elevations. It develops best in moist soils and is a frequent component of meadows, old pastures, and newer irrigated pastures, where it is an excellent forage grass. It seldom persists for long on drier soils, although it is rather common along roads or in roadside ditches where moisture conditions are better for its survival. Perennial Ryegrass is often used in lawn mixtures since it rapidly produces a cover and is much finer leaved than *Lolium multiflorum*.

3. *Lolium temulentum* (fig. 5) Darnel; Darnel Ryegrass

Coarse annual; culms 30–90 cm. tall; spike stiff; spikelets 5–7-flowered, the florets becoming plump as the grain matures; lemmas in most plants awned, though, in one form of the species, no awn is produced; glume often exceeding all of the florets.

Introduced from Europe. Now naturalized and occasional at low elevations throughout much of California. Darnel often develops well on disturbed soils but fares rather poorly on rich soils where there is much established and often varied competition. It seldom occurs in dense stands; hence it is unimportant as a forage grass. The plump grains are sometimes subject to a fungal infection, ergot, which eventually produces a visibly black sclerotium in place of the grain. Ergot is poisonous to animals and humans and can be fatal, so the grass is sometimes known as Poison Darnel.

3. Genus *Poa* Bluegrass

Perennials, or some annuals; blades flat or folded to occasionally slender and involute, the margins folded together at the tip and resembling the prow of a boat; panicles open to contracted; spikelets 2–8-flowered, mostly flattened but sometimes subcylindric; glumes somewhat unequal, the lower 1-nerved, the upper 3-nerved; florets awnless, readily separating from each other by disjointing of the rachilla; lemmas 5–nerved, variously hairy or scabrous, rarely entirely smooth, commonly scarious about the apex.

A large genus of over 200 species throughout the temperate regions of the world and high mountain chains in the tropics. There are 36 species in California many of which are valuable as forage for livestock, in soil stabilization, and as turf (in particular, *Poa pratensis*). As range grasses the tufted species occur mostly on dry soils of slopes, ridges or flats from the seacoast, foothills, and mountains, to the semi-arid regions. Most of the rhizomatous species prefer moist soils and

are common in or about meadows in the mountains.
1. Annual ....1. *P. annua*
1a. Perennial.
    2. Plants with rhizomes.
        3. Culms wiry, noticeably 2-edged; lemmas with no long cottony web at base........2. *P. compressa*
        3a. Culms not especially wiry, mostly cylindrical; lemmas with a long, cottony web at base...........
            3. *P. pratensis*
    2a. Plants without rhizomes.
        4. Lemmas ordinarily with some crisp, short hairs toward base; blades soft............
            4. *P. scabrella*
        4a. Lemmas uniformly scaberulous without short hairs toward base; blades firm............
            5. *P. nevadensis*

1. *Poa annua* (fig. 13a) Annual Bluegrass

Bright green annual; culms 5–30 cm. tall; blades flat, somewhat crinkled about the middle; panicles somewhat pyramidal in outline, 3–8 cm. long; spikelets 3–6 mm. long; lemma short-hairy on the nerves, the lateral nerves hairy only toward the base.

Naturalized throughout California from sea level to high elevations and growing wherever there is sufficient moisture. It is the earliest flowering grass of low elevations and sometimes here produces a second crop of plants later in the same year. Annual Bluegrass is a common spring weed in cities and towns. On open rangelands, any good stand of the grass is valuable as sheep forage.

Fig. 13. Florets of the genus *Poa*; a, *Poa annua*; b, *Poa scabrella*; c, *Poa pratensis*

2. *Poa compressa* Canada Bluegrass

Rhizomatous blue-green perennial; culms wiry, noticeably flattened, 15–45 cm. tall; panicles usually rather compact and short, 3–8 cm. tall; spikelets subsessile on the short branches and 4–6 mm. long, 3–6-flowered; lemmas 2–3 mm. long, hairy on the margin and midrib in the lower half, with a small weblike tuft at the very base of the midrib.

Naturalized (although perhaps some forms are native) in

California, along the north coast and medium to high elevations in the North Coast Ranges, Cascades, and Sierra Nevada, primarily in the northern half of the state. Canada Bluegrass is sometimes used as a turfgrass or as pasture on rather poor clay soils, where it thrives, but it is unable to compete with other grasses on rich, deep, and loamy soils.

3. *Poa pratensis* (pl. 2e, fig. 13c) Kentucky Bluegrass

Perennial by slender, rather widely creeping rhizomes; culms 30–100 cm. tall; blades flat or folded, 10–30 cm. long; panicles 6–12 cm. long, open and pyramidal, the lower branches ordinarily in a whorl of 5 and spikelet-bearing toward the ends; spikelets 3–5 mm. long, lemmas 3–3.5 mm. long with a very long tuft of cottony hairs (web) at the base and more or less silky-hairy on the keel and marginal nerves below.

Throughout California and North America with some forms probably native, the others naturalized from Europe. Kentucky Bluegrass is extremely important in lawns either in pure stand or in mixture with other grasses. It is a major species of mountain meadows, where it is grazed by livestock, it tolerates close grazing and much trampling, easily recovering by virtue of its extensive underground rhizomes.

4. *Poa scabrella* (pl. 2f, fig. 13b) Pine Bluegrass

Tufted perennial forming large or small tufts with soft basal foliage that in many areas withers upon or shortly after flowering; culms rather stout, 30–80 cm. tall; panicles densely or loosely contracted, 5–15 cm. long; spikelets 6–10 mm. long, pale green to purplish; lemma 3–5 mm. long, more or less rounded on the back, commonly short, crisp-hairy about the base.

Native. Throughout California from sea level to timberline and into the deserts. At low elevations the grass becomes dormant with the approach of the hot, dry summer; in the cool mountain regions it thrives from the time of snow melt in the spring to late in the summer. Domestic animals graze the tender foliage throughout the summer at high elevations. Pine Bluegrass, at one time, was quite abundant in the Central Valley along with *Stipa pulchra* and other species.

5. *Poa nevadensis* (pl. 2g) Nevada Bluegrass

Tufted perennial of small to large bunches; culms and foliage uniformly scabrous; culms 45–75 cm. tall; blades rather stiffly ascending, not at all withering as in *Poa scabrella* (with which it could be confused); panicles loosely contracted, as much as 25 cm. long, mostly pale green; spikelets 4–8 mm. long; lemmas 3.5–5 mm. long, mostly finely scabrous over the entire back.

Native. Mostly above 4000 feet, dry to moist mountain meadows or occasionally rocky slopes; occurs from Inyo County north along the east side of the Sierra Nevada and Cascades to Modoc County, thence north to Washington and east to the Rocky Mountains, south to Arizona. It is a valuable forage grass of the summer rangelands and produces much succulent forage early in the season before flowering.

4. Genus *Dactylis*
1. *Dactylis glomerata* (fig. 14) Orchardgrass

Perennial, often forming large clumps, the young shoots strongly flattened and whitish; blades V-shaped in cross section, 8–20 cm. long; ligule a prominent membrane; panicles 10–25 cm. long, the lower branches stiffly spreading and bearing dense, ovoid clusters of strongly flattened spikelets at the ends; spikelets several-flowered; glumes unequal, rough-hairy on the keel; lemmas about 8 mm. long, flattened, slightly curved, tapering into a short point at the apex.

Fig. 14. *Dactylis glomerata* panicle

Introduced from Europe. Now naturalized. Valuable as a hay and pasture grass on irrigated soils; on rangelands it is best adapted to moist soils in the mountains. Stands of the grass are relatively sparse probably because of excessive grazing by livestock and deer and damage by rodents. Orchardgrass does not reseed itself well except on disturbed soils. Seeding the grass in mountain meadows and on burned-over or logged forests has met with some success in certain areas.

5. Genus *Briza* Quakinggrass

Annual (California species) herbage pale green; panicles open, delicate, the flattened spikelets pendulous on hairlike pedicels and fluttering in the wind (hence the common name); glumes and lemmas quite broad, papery, more or less horizontally disposed.

1. Spikelets numerous, about 3 mm. long........1. *B. minor*
1a. Spikelets relatively few, 10–20 mm. long....2. *B. maxima*

1. *Briza minor* (fig. 15) Little Quakinggrass

Culms 7–35 cm. tall; panicles more or less pyramidal in outline, diffuse; spikelets somewhat triangular in outline, about 3 mm. long, 3–6-flowered; lemmas smooth.

Introduced from Europe. Now naturalized throughout most of California at lower elevations and commonly found in low places in the valleys or about the base of foothills, especially in

Fig. 15. *Briza minor* spikelet and habit

northern California where the annual rainfall is greater. Because of its usually low stature, Little Quakinggrass is sometimes concealed among taller grasses such as Wild Oats or Annual Ryegrass, but on less fertile soils it may be quite conspicuous. On the valley and foothill rangelands it affords some forage for sheep before flowering, after which it is valueless, although the delicate panicles are used in small, dry bouquets. The plants are sometimes grown in gardens where they are of ornamental value.

2. *Briza maxima* (pl. 2h) Big Quakinggrass

Culms 20–60 cm. tall; panicles reduced to a few large, long-pedicelled spikelets 10–20 mm. long; lemmas more or less closely short-hairy.

Naturalized mostly along the California coast but it does occur in the foothills of the Coast Ranges and northern Sierra Nevada. The panicles are attractive in dry arrangements though the florets readily shatter if collected too late. The grass is quite ornamental growing in gardens. As a livestock forage it is poor.

Fig. 16. *Lamarckia aurea*; a fascicle of spikelets

6. Genus *Lamarckia* Goldentop
1. *Lamarckia aurea* (fig. 16) Goldentop

Annual; culms 7–40 cm. tall; foliage pale to yellowish-green, blades flat, soft; ligules prominent, membranous; panicles commonly yellowish, 1-sided, oblong, 2.5–10 cm. long, consisting of numerous fascicles of spikelets, each fascicle mostly of sterile spikelets surrounding a single fertile spikelet of 2 florets, these florets long-awned. The fascicles readily fall away as units from the inflorescence and by this character the grass is readily identified.

Introduced from Europe. Naturalized at low elevations, mostly in the Coast Ranges from Mendocino County south to San Diego County, occasionally in the Sierra Nevada and southern San Joaquin Valley. Goldentop is an attractive and unusual grass having been grown as an ornamental in gardens from which it probably escaped. It is a poor competitor among other grasses or other vegetation. Commonly it is found in areas of disturbed, thin, or even rocky soils. Abundance of the grass on rangelands indicates poor condition of that range since the species is inferior forage.

7. Genus *Cynosurus* Dogtail

1. *Cynosurus echinatus* (fig. 17) Hedgehog Dogtail

Annual; culms 10–50 cm. tall; herbage pale green; panicles ovoid to oblong, dense and bristly, placed on one side of the stem at the apex, 1–4 cm. long; spikelets of two kinds in the same panicle, sterile and fertile intermixed; fertile spikelet 2–3-flowered, the florets long-awned and readily falling from the glumes; sterile spikelet of several empty lemmas that are attenuate into an awn at the tip.

Fig. 17. *Cynosurus echinatus*; habit and the two kinds of spikelets

Introduced from Europe. Naturalized at low elevations mostly below 3000 feet in the Coast Ranges and Sierra Nevada. Hedgehog Dogtail is of little value as a forage plant and its abundance in some areas is of concern since it indicates range deterioration. It matures later than most annual grasses, so early and heavy grazing removes the choicer grass species and allows Hedgehog Dogtail to develop and mature a heavier seed crop because of less competition. It frequently occurs in dense stands under oaks in the foothills.

The panicles are decorative in dried arrangements. There is some shattering of the fertile florets but this does not affect the bristly appearance much.

8. Genus *Melica* Melic; Oniongrass

Tufted perennials; culms in some species with swollen bases and these are known commonly as Oniongrasses; foliage scaberulous, sheaths tubular; panicles open, contracted, to dense and spikelike; spikelets with 1–several readily deciduous grain-producing florets, the uppermost floret (rudiment) sterile, consisting really of several sterile lemmas closely wrapped around each other to form a knob; glumes ordinarily subequal, much like parchment or papyrus in texture; lemmas rounded on the back, conspicuously 5 to many-nerved, commonly scabrous and sometimes pubescent as well, membranous to leathery in texture, the apex and margins usually papery, commonly with a purplish band across the back below the apex; grain grooved along one side, sometimes rather loosely enclosed between the lemma and palea.

Twelve species are found in California of which the three described below are probably the most common. The Melics occur on dry soils and seldom, if ever, are found in dense stands. They are of minor importance as forage even though grazed. Melics are of some value as soil binders especially on roadcuts through the foothill regions and on disturbed brushland soils.

1. Spikelets large, 10–20 mm. long, whitish to purplish; culm bases somewhat thickened but not at all bulbous........
1. *M. californica*
1a. Spikelets smaller, 5–6 mm. long, commonly purplish; culm

bases not at all thickened.
2. Stipe of the rudiment about 0.5 mm. long; glumes not sharply acute....................2. *M. imperfecta*
2a. Stipe of the rudiment about 2.5 mm. long; glumes sharply acute......................3. *M. torreyana*

1. *Melica californica* (pl. 3a) California Melic

Densely to loosely tufted perennial; culms 50–100 cm. tall, the bases somewhat enlarged but scarcely bulbous; panicles contracted, narrow, whitish, sometimes purplish; spikelets large, 10–20 mm. long with 2–4 grain-producing florets beside the rudimentary one; lemmas obviously parchmentlike in texture, 7-nerved, awnless.

Native. Slopes and canyons of the foothills, oak woodland, occasionally in brushlands and pine forests at lower elevations, in the Sierra Nevada and Coast Ranges north to southwestern Oregon. California Melic, *Stipa pulchra*, and *Poa scabrella* are characteristic species of the foothill rangeland. This Melic is fairly good forage for cattle, and sheep graze the young foliage early in the season.

2. *Melica imperfecta* (fig. 18a) Coastrange Melic

Densely or sometimes loosely tufted perennials; culms 30–90 cm. tall; panicle extremely variable as to its length and disposition of

Fig. 18. Spikelets of two *Melica* species; a, *Melica imperfecta*; b, *Melica torreyana*

the fascicled, unequal branches (closely appressed to widely spreading or even reflexed); panicle 10–30 cm. long; spikelets 5–6 mm. long, commonly purplish, of 1 or 2 fertile florets, the oblong rudiment above on a short pedicel about 0.5 mm. long; glumes rather obtuse at the apex.

Native. At low to medium elevations in the Coast Ranges from San Diego County north to Lake County; also central and southern Sierra Nevada and western edge of the Mojave Desert. Coast-range Melic is commonly associated with brushlands or rocky soils in the oak woodland and lower elevation coniferous forest. It is a valuable forage grass for all classes of livestock; the early growth is the most palatable and succulent. Good stands of this Melic have been greatly reduced by overgrazing.

3. *Melica torreyana* (cover-A, fig. 18b) Torrey Melic

Similar in growth habit to *Melica imperfecta* but occurring mostly in the North Coast Ranges and in the northern Sierra Nevada. It differs by the usually narrow, contracted panicle, longer and rather sharply acute glumes, and the rudiment borne on a pedicel about 2.5 mm. long.

Native. Common in brushlands particularly on serpentine soils. Following brush fires this Melic and *Stipa lepida*, a companion species, become very abundant; their density gradually decreases with increasing competition from other plants. It is similar to *Melica imperfecta* as range forage.

9. Genus *Glyceria* Mannagrass

Perennials of moist or wet habitats with creeping and rooting culm bases, or rhizomatous; stems and foliage finely scabrous; blades flat; panicle open; glumes thin, rather minute in some species; lemma scabrous, rounded on the back, 5–9-nerved, the nerves prominent, raised, parallel with the midnerve, the apex obtuse and scarious.

1. Sheaths split; culms less than 80 cm. tall...1. *G. pauciflora*
1a. Sheaths closed; culms commonly 1 meter or more tall....
   2. *G. elata*

1. *Glyceria pauciflora* (fig. 19) Weak Mannagrass

Rhizomatous perennial; culms 40–80 cm. tall; foliage green or sometimes purplish, sheaths split, blades flat 5–15 mm. wide, 8–15 cm. long; panicles open, 10–15 cm. long, the branches more or less wavy or flexuous; spikelets commonly purplish, sometimes pale green, 4–5 mm. long, 5- or 6-flowered; glumes thin, the lower one 1–1.5 mm. long, the upper one much longer and 3-nerved; lemmas with 5 conspicuous parallel nerves, the lemma apex rather thin and transparent.

Native. Wet slopes, meadows, and bogs, at medium to high

elevations in the mountains from southern California north through the Sierra Nevada and Cascades; North Coast Ranges and along the north coast. Weak Mannagrass is well grazed by livestock because of its succulence and occurrence in or about water where the animals drink. In many areas the grass has been grazed out or the stand greatly reduced to scattered plants. Despite grazing pressure on this species it is still the most abundant of the Mannagrasses in California. In its natural setting the grass is very attractive because of its broad foliage and delicate panicle.

2. *Glyceria elata* (fig. 20) Tall Mannagrass

Rhizomatous perennial; culms 100–150 cm. tall; sheaths closed, blades flat, 5–15 mm. wide; panicles open, 15–30 cm. long, the branches wide-spreading, nodding; spikelets 4–6 mm. long, 6–8-flowered; glumes unequal, minute, both less than 1.5 mm. long, the upper one only 1-nerved; lemmas with 7 prominent parallel nerves and 2–2.5 mm. long.

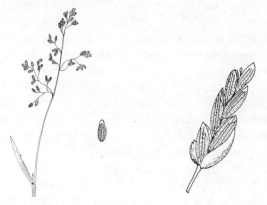

Fig. 19. *Glyceria pauciflora*; habit and the back of a single floret
Fig. 20. *Glyceria elata* spikelet

Native. In or around fresh water streams, lake margins, meadows, bogs and swamps, mostly at medium elevations but ascending to about 8500 feet: Cascades, Sierra Nevada, south to the higher mountains of southern California; North Coast Ranges and occasional along the north coast of California at or near sea level. This Mannagrass is attractive because of its tall stems and the graceful nodding branches of the open panicle. As forage it is very palatable and is well grazed early in the season.

10. Genus *Bromus* Brome; Chess

Annuals or perennials; foliage of most species more or

less hairy, sheaths closed by fusion of the margins; panicles open or contracted, if dense then some lower branches well developed; spikelets large, strongly flattened to sub-cylindric, several- to many-flowered, the florets readily shattering, the glumes persistent; lemmas 5–13-nerved, strongly keeled to rounded on the back, variously hairy, scabrous or sometimes smooth, commonly awned from or slightly below the apex; palea commonly hairy along the margins; grain flattish, concavo-convex in cross section, hairy at the apex.

A large genus of more than 100 species it grows best in the temperate zones of the world. California alone has 36 species, over half of them introduced from the Mediterranean region. All of the perennials and some of the annuals are valuable as livestock forage either as pasture or hay. The seed heads, excepting those few with rigid awns and sharp-pointed florets, are nutritious and frequently grazed along with the stems and foliage.

The Bromes prefer the drier soils of the foothills and mountains though a few, such as *Bromus inermis*, do well under meadow conditions. The annual species are aggressive and pioneer on disturbed soils but many persist under strong competition. The annual Bromes, especially *Bromus mollis*, are among the most important forage species on the winter annual California rangelands. A few such as *Bromus tectorum* and *Bromus rubens*, are significant grasses in the Great Basin and Mojave Desert. They are of value as forage only during the period of vegetative development; upon formation of the inflorescence they lose their appeal to grazing animals. The annuals are common spring weeds along roads, fencerows, vacant lots in towns, in orchards, vineyards, crop lands, and various other habitats. At low elevations in the state these grasses are mature by May or June depending upon the latitude. Those species with stiff and long awns and sharp-pointed florets, such as *Bromus diandrus*, are bothersome to animals because they enter ears, nostrils, and sometimes eyes, causing severe irritation or even inflammation. The florets are annoying in animal hair and man's clothing.

1. Perennial.
    2. Spikelets strongly flattened; lemmas V-shaped in cross section, keeled the entire length, more or less uniformly scabrous or closely short-hairy over the back.
        3. Lemmas with awns in excess of 5 mm. in length..
            1. *B. carinatus*
            2. *B. marginatus*
        3a. Lemmas awnless or at most short-awned, the awn not exceeding 3 mm. in length.
            4. Lemmas 11–13-nerved.....3. *B. willdenowii*
            4a. Lemmas 7–9-nerved........4. *B. unioloides*
    2a. Spikelets not strongly flattened, the lemmas more or less rounded on the back though may be somewhat keeled towards the apex.
        5. Plants strictly tufted; lemmas obviously hairy at least in the lower half with longer hairs along the margin..5. *B. laevipes*
        5a. Plants with short creeping rhizomes about the base; lemmas smooth or at least not obviously hairy..6. *B. inermis*
1a. Annual.
    6. Lemmas relatively broad (if folded out flat would generally be elliptic in outline), shallowly single-notched at the rounded apex, the midnerve prolonged as a straight or divergent awn from the sinus; callus blunt.
        7. Panicle dense, compact, of few to many spikelets, the branches short and always erect or ascending when in fruit.................................7. *B. mollis*
        7a. Panicle open, the long branches flexuous or drooping in fruit.
            8. Lemmas hairy, the pedicels and sometimes the branches flexuous...............8. *B. arenarius*
            8a. Lemmas without hairs, the branches and pedicels not at all flexuous but rather are drooping in fruit.
                9. *B. commutatus*
    6a. Lemmas scabrous, narrow and elongate, the apex with two bristlelike teeth 2–5 mm. long; callus sharp-pointed.
        9. Panicles open or sometimes loose or narrow

but at least with some long branches that are curved or droop in fruit.

    10. Upper glume ordinarily less than 1 cm. long; pedicels and even some of the branches flexuous; lemma awn 15 mm. or less long.

        10. *B. tectorum*

    10a. Upper glume longer than 1 cm.; lemma awn 2 cm. or more long.

11. Lemma 20 mm. or less long; lemma awn about 2 cm. long; lower glume 8 mm. long..................11. *B. sterilis*

11a. Lemma 25 mm. or more long; lemma awn 3–5 cm. long; lower glume 15 mm. long..............12. *B. diandrus*

    9a. Panicles more or less compact, dense and somewhat ovoid in outline, the branches relatively short, erect or sometimes spreading but not at all long nor curved and drooping in fruit.

    12. Spikelets densely aggregated, the short branches erect or ascending; culms ordinarily densely short-pubescent below the panicle....................13. *B. rubens*

    12a. Spikelets loosely aggregated, the branches somewhat spreading; culms ordinarily glabrous below the panicle.

        14. *B. madritensis*

1. *Bromus carinatus* (fig. 21) California Brome

Short-lived perennial or some forms annual; culms 50–120 cm. tall; foliage variously hairy to glabrous; panicles open, 15–30 cm. long, the lower branches usually elongate and eventually drooping; spikelets strongly flattened, ordinarily 2–3 cm. long, but in some as much as 5 cm. long (excluding the awns), 5–many-flowered; lemmas strongly keeled, more or less uniformly scabrous or short-hairy over the sides, 10–20 mm. long tipped with a straight awn 7–15 mm. long.

Native. Mostly in wooded habitats of the foothills and mountains, distributed throughout the state; sometimes in dense stands in the open pine or oak forests where it provides good forage. The young growth is nutritious and readily grazed by animals, but upon elongation of the stems it becomes less palatable.

2. *Bromus marginatus* Mountain Brome

Similar to and perhaps not specifically distinct from *Bromus carinatus*. It does, however, have a much stricter panicle and

Fig. 21. *Bromus carinatus*; a, habit; b, spikelet; c, floret

much shorter lower branches. Ordinarily the sheaths are densely hairy and the blades quite narrow.

Native. Widely distributed over the state, primarily in the mountains where it is a valuable forage during the summer period; sometimes abundant locally but more often dispersed over a wide area. It soon occupies disturbed soils, often quite densely so, but gradually declines over succeeding years with increasing competition from other plants.

3. *Bromus willdenowii* (pl. 3b) Prairie Brome
Short-lived perennial or some forms annual; culms 60–100 cm.

tall; foliage variously pubescent; panicles open, the lower branches as much as 15 cm. long, all of the branches drooping in fruit; spikelets strongly flattened, 6–12-flowered, 2–3 cm. long; lemmas short-hairy to scabrous, 11–13-nerved, the apex acute or with a short awn 1–3 mm. long.

Introduced from South America. Now naturalized mostly on deep, rich soils of valleys. It is a valuable pasture grass with strong, vigorous growth during the winter, and it flowers early in spring. The seed heads are nutritious and are avidly grazed by dairy cows and cattle with the foliage. Prairie Brome, owing to its short life, does not satisfactorily persist in irrigated pastures and must be planted with other grasses and legumes of longer duration or better reseeding characteristics. It is often found along roads and roadside ditches, on land fills, and sometimes in cities and towns.

4. *Bromus unioloides* Rescue Brome

Similar to *Bromus willdenowii*, (syn. *Bromus haenkeanus*) differing only in the 7–9-nerved lemmas.

Apparently this species is commoner in southern California.

5. *Bromus laevipes* Chinook Brome; Woodland Brome

Perennial forming tufts; unlike most bromes, the foliage is entirely glabrous; panicles 8–15 cm. long, the branches slender and eventually drooping in fruit; spikelets 1.5–4 cm. long, 5–13-flowered; lemmas hairy with longer hairs along the margin in the lower half, rounded on the back in the lower half and more or less keeled above.

Native. Occurs on wooded slopes at low to medium elevations; Coast Ranges from San Luis Obispo County north; west slope of the Sierra Nevada; thence north to Washington. Chinook Brome is valuable in holding soil, particularly in shaded roadcuts of the foothill regions and lower mountain slopes. It seldom occurs in dense stands, but when it does livestock graze the young foliage with relish.

6. *Bromus inermis* Smooth Brome

Perennial by rhizomes; culms 50–120 cm. tall; the basal sheaths becoming fibrous and brownish in age, blades glabrous, 5–10 mm. wide; panicles open or sometimes rather loosely contracted, 15–25 cm. long; spikelets 20–25 mm. long; lemmas smooth or scabrous, sometimes hairy on the lower half, the apex more or less obtuse, usually awnless but sometimes with a short awn 1–2 mm. long.

Introduced from Eurasia. Now somewhat naturalized at medium elevations in the mountains of California. It requires a favorable seed bed to obtain a good stand and is frequently used in seedings for pasture improvement in the mountains. Smooth Brome is valuable because of its excellent production of leaves, which are useful as hay or pasture, and its strong rhizome system which allows close utilization of the herbage and greater recovery. Its tolerance of cold and dryness is advantageous in its establishment and persistence. The grass is not adapted to lower elevations because of the hot and dry summers.

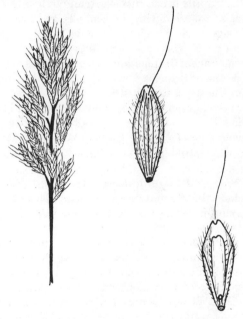

Fig. 22. *Bromus mollis*; panicle and a back and front view of a floret

7. *Bromus mollis* (fig. 22) Soft Chess; Soft Brome

Annual; culms 10–40 cm. tall, in robust specimens to nearly 80 cm. tall; sheaths densely soft-hairy, the blades less densely hairy; panicles dense, compact, 4–10 cm. long ordinarily of several to many spikelets but in spindly plants may be reduced to 1 or 2 spikelets; spikelets 12–20 mm. long, 6–12-flowered; lemmas rounded on the back and on the callus, softly hairy, or in the forma *leiostachys*, smooth, the awn soft 5–10 mm. long, straight or curved to one side.

Introduced from Europe. Now naturalized and common at low elevations throughout most of California. Soft Chess is one of the most abundant and important forage species on the valley and foothill rangelands and is grazed early through maturity since the seed heads are very nutritious. It performs well when seeded on brush burns either sown directly in the ash or drilled into the intervening spaces. The species is a common weed about cities, towns, along roads, ditches, fencerows, trails, etc.

8. *Bromus arenarius* Australian Brome

Annual; culms 15–40 cm. tall, slender, solitary or several from the base; foliage hairy; panicles open, 8–11 (–15) cm. long, the branches and pedicels slender, commonly S-shaped or at least curved; lemmas about 10 mm. long, tipped with a slender awn as long as or longer than the lemma.

Australian Chess is naturalized throughout much of the state and occurs primarily on sandy or light soils below about 5000 feet elevation. It is good forage and is often abundant on soils which ordinarily support little other, or at least poor quality, forage. Any dense stand of the grass tends to reduce soil erosion.

9. *Bromus commutatus* Hairy Brome

Annual; culms 30–100 cm. tall; sheaths and blades hairy; panicles loose, mostly open, the long branches eventually drooping more or less to one side of the rachis and bearing spikelets towards the ends; spikelets 15–20 mm. long; lemmas 7–11 mm. long, more or less rounded on the back, smooth to finely scabrous, the awn straight, 4–10 mm. long.

Naturalized mostly in meadows, moist slopes, along streams or in swales: Coast Ranges, Sierra Nevada, and northern counties of the state. The grass is of some value as forage for livestock where abundant.

10. *Bromus tectorum* (pl. 3c) Cheatgrass Brome; Downy Brome

Annual; culms 10–40 cm. tall; basal foliage hairy but becoming less dense on the upper leaves; panicles open or at least loose, 5–15 cm. long, the branches and pedicels hairlike, curved or often S-shaped; florets slender, smooth, scabrous to hairy, curved;

lemmas 8–12 mm. long tipped by a longer awn.

Naturalized throughout California at all elevations. Occurs on poor or sandy soils where other vegetation does not offer too much competition, so it is best developed in the deserts, particularly the Great Basin, where it is a common associate of sagebrush (*Artemisia tridentata*). In this region absence or sparse stands of bunchgrasses favor Cheatgrass Brome, and it soon assumes dominance. Early in the season it is grazed by livestòck, but upon elongation of the stems and emergence of the inflorescence it loses its appeal to the animals. Where bunchgrasses are the primary forage, an abundance of Cheatgrass Brome indicates range deterioration. On rich soils and in less arid conditions the grass is a poor competitor among other plants. It is an occasional weed along roads or in cities and towns.

The common name, Cheatgrass, as applied to this Brome species refers to its abundance as a weed in areas where cereal grains had been planted many years ago.

11. *Bromus sterilis* Poverty Brome

Annual; similar in some respects to *Bromus diandrus*, mostly rather delicate, the panicle 10–20 cm. long with long drooping branches and the lemmas 17–20 mm. long.

Introduced from Europe. Now naturalized at low elevations throughout much of California. Flowers early in the spring and shatters away long before other annual grasses have matured. It has narrower and shorter leaves than does *Bromus diandrus* and is of much less value as forage for livestock. Poverty Brome is sometimes abundant locally but more often is spread over wide areas without being common in any one place. It varies from frequent to occasional as a weed along roads, waste areas, or in cities and towns.

12. *Bromus diandrus* (pl. 3d) Ripgut Brome

Annual; culms 30–70 cm. tall; sheaths and the flat blades hairy and rather coarsely so; ligule 3–4 mm. long, jagged along the margin; panicles 6–18 cm. long, the lower branches from 1–5 cm. long; spikelets 3–5 cm. long (excluding the awns); lemmas scabrous, 2.5–3 cm. long tipped by a scabrous awn 3.5–5 cm. long, the callus at the base rounded.

Bromus rigidus, a native of Europe to which our Californian material has always been referred, differs in having a needle-sharp callus at the base of the lemma, which none of the California species examined has. The common name, Ripgut Brome, is here applied to *Bromus diandrus*.

Introduced from Europe. Now naturalized throughout California at low to medium elevations; common on valley flats and foothill slopes. Ripgut Brome is good forage when young, providing much succulent and nutritious grazing, but it matures early and the long awns and sharp florets are injurious to the mouth, nostrils, ears, and sometimes eyes of animals.

Ripgut Brome is a common weed of cities and towns, where it occurs in vacant lots or along sidewalks, and it is also abundant along roads and fence rows, and in open fields. It grows well on nearly all soils from adobe clay to sand and is quick to invade disturbed sites.

13. *Bromus rubens* (pl. 1f, 3 e) Red Brome; Foxtail Brome

Annual; culms 15–40 cm. tall; finely close-hairy on the leaves and on the culm below the panicle; panicles dense, compact, ordinarily ovate in outline, 2–7 cm. long; spikelets rather closely appressed or ascending, 7–11-flowered, about 2.5 cm. long, the dark-reddish florets curved outward at maturity; lemmas scabrous to hairy, 12–16 mm. long and tipped by a longer awn.

Introduced from Europe. Now naturalized throughout California at low to medium elevations, common especially where competition is at a minimum, as in the Mojave and Colorado deserts, the arid hills and flats on the west side of the San Joaquin Valley, or the thin and stony soils of brushlands in the foothills where it is valuable in preventing soil erosion when in good stand. Ordinarily it is poor forage because of its sparse foliage and early maturity. The rather stiff awns and sharp-pointed florets are objectionable to grazing animals. In other areas, besides the deserts and brushlands, Red Brome is a common or occasional weed in or about cities and towns, along roads, trails, disturbed soils, overgrazed foothill ranges, etc. Strong competition from other plants often greatly reduces the density of the grass.

14. *Bromus madritensis* Spanish Brome; Madrid Brome

Annual, similar to *Bromus rubens* and, in depauperate forms, scarcely distinguishable from that species. Ordinarily the panicle is loose and the spikelets more or less divergent from the rachis. The culm is ordinarily without hairs below the panicle.

Introduced from Europe. Now naturalized and best developed on thin or rocky soils of brush areas in California. Here it is a pioneer on disturbed soils, at first forming a dense stand but in succeeding years gradually diminishing in abundance.

Fig. 23. *Brachypodium distachyon*; spikelet and habit

11. Genus *Brachypodium* Falsebrome
1. *Brachypodium distachyon* (fig. 23)
Purple Falsebrome

Annual; culms 10-30 cm. tall, rather wiry, hairy at the nodes; foliage sparsely hairy; spikelets large, 1 to 5 in number, sessile, or some provided with a very short pedicel, aggregated about the apex of the culm; spikelets 20-25 mm. long, 10-15-flowered, the rachilla quite tardily disjointing, sometimes not at all until fall; lemmas 8-9 mm. long, rounded on the back, scabrous and tapering into a straight, rigid awn 1-2 cm. long; palea with stiff hairs along the margin in the upper half.

Introduced from Europe. Now naturalized and rather generally spread over California at low elevations; sometimes locally abundant on thin rocky soils, at the edges of a dense shrub cover, disturbed soils, and open foothill slopes. It is poor forage, scarcely palatable to livestock because of the wiry culms, sparse foliage, and stiff-awned seed heads.

12. Genus *Triticum* Wheat
1. *Triticum aestivum* (pl. 3f) Wheat

Annual; culms stout, 50–100 cm. tall; blades flat, exceeding 1 cm. in width, auriculate at the base, the auricles clawlike; spike thick, 5–15 cm. long; spikelets flattened, the flat side placed against the rachis, several-flowered; glumes tough with a strong midrib that is somewhat skewed to one side and ending abruptly as a short point at the apex; lemmas several-nerved, the midrib off-center and ending in an abrupt point or a conspicuous long awn.

A cultivated crop in California, occasionally volunteering along roads.

13. Genus *Aegilops* Goatgrass
1. *Aegilops triuncialis* (pl. 3g) Barb Goatgrass

Annual; culms 20–40 cm. tall and usually stiffly erect; spike nearly cylindrical, of 3–5 large spikelets; glumes very tough, each ending in three stiff, stout and eventually spreading long awns. At maturity the whole spike falls away to the ground as a unit and remains so until the soaking rains in the fall breaks it up into joints. The seed germinates in a portion of the spike, and that blackish portion may be found among the roots of the barb goatgrass plant.

Naturalized over much of cismontane California; locally common in the Coast Ranges, Great Valley, and western slope of the Sierra Nevada. Barb Goatgrass is a serious weed of the foothill rangelands where it often forms dense and extensive patches thereby crowding out desirable forage species. Livestock usually avoid the grass because of its stiff stems, scarcity of foliage and the obnoxious spikes. It invades rangelands that have been continuously overgrazed or may establish itself on disturbed often infertile soils. The density of the Goatgrass species may be greatly reduced by seeding infestations with annual Clovers and Hardinggrass and applying fertilizer.

The genus *Aegilops* is closely related to wheat (*Triticum*) and species of both *Aegilops* and *Triticum* have contributed to the genetic makeup of this valuable cereal crop.

14. Genus *Hordeum* Barley

Annuals or some perennials; spikes bristly in most by long slender awns and glumes; spikelets in 3's side-by-side at each node of a flattened rachis, the slender glumes standing in front of each spikelet; central spikelet ordinarily sessile on the rachis and maturing a single grain, the lateral pair of spikelets mostly pedicelled and rudimentary or sometimes well developed and staminate; triads of spikelets are usually numerous, alternate on opposite sides of the flattened rachis; at maturity the rachis readily separating at the joints and the triad of spikelets falling entire with an internode of the rachis attached below; lowermost 2 or 3 triads of spikelets often completely sterile and remain attached to the tip of the stem; deciduous segments of the spike are often thickly deposited upon the ground and these readily identify the genus.

The annual species are common at low elevations throughout California. Some, such as *Hordeum geniculatum* and *Hordeum depressum*, are common over the broad alkaline flats in the Central Valley, while *Hordeum leporinum* and *Hordeum glaucum* are abundant on foothill slopes as well as common weeds in cities and towns, along roads, fencerows, in pastures, croplands and generally in waste areas. As range grasses only the early growth of the plant is of value as forage, for upon flowering the bristly spikes become objectionable to grazing animals. The rough, barbed triad of spikelets are a great nuisance to animals since they become entangled in hair or enter nostrils, ears, eyes, and throat. The upwardly directed barbs on the spikelets make withdrawal of the triad difficult.

The perennial species are far less common but in local areas or certain habitats may be of sufficient density to be valuable as forage or, in the case of *Hordeum jubatum*, obnoxious as a weed.

Hordeum vulgare, cultivated barley, differs from all of the other species by the nonfracturing rachis, the spike remaining intact awaiting harvest.

1. Perennial.
    2. Spike about as broad as long, very bristly with awns

       2–5 cm. long......................1. *H. jubatum*
   2a. Spike much longer than broad, not especially bristly,
       the awns 1 cm. or less long......2. *H. brachyantherum*
1a. Annual.
       3. Auricles at the base of the blade well developed,
          clawlike.
          4. Rachis readily disjointing at the nodes, cen-
             tral spikelet (as well as the lateral) pedi-
             celled; glumes of the central spikelet and
             inner one of the lateral spikelets ciliate......
                                              3. *H. leporinum*
                                              4. *H. glaucum*
          4a. Rachis not disjointing; spikelets all sessile;
              glumes not ciliate............5. *H. vulgare*
       3a. Auricles wanting.
             5. Floret of lateral spikelets awnless; foliage
                rather uniformly finely short-hairy (al-
                most frosty in appearance); spike 1.5–3
                cm. long............6. *H. geniculatum*
             5a. Floret of lateral spikelets awned; foliage
                 not uniformly hairy, the sheaths more or
                 less glabrous; spike 3–6 cm. long......
                                              7. *H. depressum*

1. *Hordeum jubatum* (fig. 24) Foxtail Barley

Tufted perennial; culms 20–60 cm. tall; blades 5 mm. or less broad, mostly scabrous, sometimes pubescent; spikes 5–10 cm. long, sometimes partially included in the upper sheath; central spikelet with a floret 6–8 mm. long tipped by a much longer hairlike awn; florets of the lateral spikelets reduced to slender awns; glumes are hairlike awns.

Native. Widely distributed through California from low to high elevations; prefers moist soils and sometimes occurs in dense stands in or about meadows; particularly common along roads, roadside ditches and alkaline fields or flats. Foxtail Barley may be grazed to some extent when young but loses its appeal to livestock upon flowering. The bristly heads and scabrous awns are annoying to eyes, ears, nostrils and mouths of animals.

2. *Hordeum brachyantherum* (fig. 25) Meadow Barley

Tufted perennial; culms 25–75 cm. tall; foliage hairy to glabrous; spikes 4.5–9 cm. long; floret of the central spikelet 5–10 mm. long tipped by an awn about as long, glumes shorter; lateral spikelets

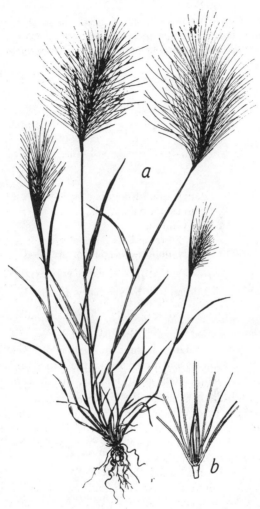

Fig. 24. *Hordeum jubatum*; a, habit of the plant; b, a joint of the spike showing the three spikelets

on rather widely out-curved pedicels.

Native. On moist soils over wide elevational distribution through California. As a forage plant Meadow Barley is of some importance in the mountain meadows where the early growth is well grazed by livestock, although its value is limited because it matures early in either high or low elevations. It is common on the mesas above the sea along the north coast of California, where the culms are often widely spreading to almost prostrate. Away from the coast the culms are only somewhat spreading or are erect.

Fig. 25. *Hordeum brachyantherum*; triad of spikelets

3. *Hordeum leporinum* (pl. 3h) Hare Barley

Annual; culms 15–60 cm. tall; foliage hairy, the flat blades with slender, clawlike auricles at the base; spike pale green to purplish, dense, 5–9 cm. long, sometimes partially enclosed at the base by the expanded upper sheath; floret of the central spikelet on a short pedicel making all three spikelets pedicelled; glumes of the central spikelet and inner one of the lateral spikelet are provided with hairs along their margin, this feature readily characterizing this species and *Hordeum glaucum*.

Introduced from Europe. Now naturalized and widespread at low elevations throughout California; common as a spring weed in and about cities and towns, along roads, fencerows, ditches, rubbish dumps, disturbed soils, and in an about croplands. On valley and foothill rangelands Hare Barley furnishes much succulent and nutritious forage when young and abundant but is of little value later because of its early maturity and rough-awned spikes.

4. *Hordeum glaucum* Glaucous Barley

Similar in most respects to *Hordeum leporinum*. Only by examination of trivial characters may this species be satisfactorily separated. It possesses the same weedy habit, forage use, and distribution in California.

Stamens of central floret are included during flowering, anthers minute, less than 0.5 mm. long; hairs on the rachis margin

[82]

are 0.25 to 0.75 mm. long; rachilla of the lateral spikelet is rather abruptly thickened at the base, the whole rachilla rather club-shaped.

In *Hordeum leporinum*, by contrast, stamens of central floret are exserted, anthers averaging about 1 mm. in length; hairs on rachis margin less than 0.5 mm. long; rachilla of lateral spikelet rather long tapering and not abruptly thickened at the base.

5. *Hordeum vulgare* (pl. 4a) Barley

Annual; culms stout, 50–100 cm. tall; blades flat with well-developed clawlike auricles at the base; spikes 5–10 cm. long, excluding the often long, stiff awns; spikelets all sessile on the rachis, the central or sometimes the lateral as well (6-row barley) grain-producing; rachis of the spike not at all breaking up into joints.

Cultivated as a crop plant; it sometimes volunteers along roads or in fields previously planted to the crop. Occasionally it is sown on roadcuts or roadslopes to prevent soil erosion.

6. *Hordeum geniculatum* (pl. 4b)
Mediterranean Barley

Annual; culms ordinarily geniculate or, in close competition, erect, 12–35 cm. tall; foliage finely short-hairy, the blades not at all auriculate; spike short, 1.5–3 cm. long; glumes rather rigid and somewhat divergent from the floret of each spikelet; lemma of the central floret about 5 mm. long, smooth and tapering at the apex into a somewhat longer awn; lateral floret awnless.

Introduced from Europe. Now naturalized at low to medium elevations throughout California, occurs in greater density on clay soils than on light or sandy soils; on the alkaline flats of the Central Valley or on the borders of marshlands it often forms dense stands. As livestock forage Mediterranean Barley is of value only early in the season, for it flowers early and the bristly head is obnoxious to grazing animals.

7. *Hordeum depressum* Alkali Barley

Annual; culms rather slender, 10–45 cm. tall geniculate or erect; foliage, especially the blades, pubescent, the sheaths more or less glabrous, uppermost sheaths commonly swollen; spikes 3–6 cm. long, erect but not stiffly so; central floret 7–8 mm. long tipped by usually a longer awn; lateral floret awnless; glumes of the three spikelets nearly equal at their tips but exceeded by the awn of the central floret.

Native. On alkaline, subalkaline or saline soils throughout the state, primarily at lower elevations. It matures early and is useful as forage only during the vegetative period. When it occurs in dense stands it is usually in association with *Hordeum geniculatum* and *Distichlis spicata*.

Alkali Barley closely resembles *Hordeum brachyantherum* in spikelet characters and is separable primarily upon its size and longevity. Considerable care is necessary to distinguish the two species.

15. Genus *Sitanion* Squirreltail

Densely tufted perennials; culms in most 15–50 cm. tall; leaves variously hairy to glabrous, blades flat to tightly rolled, often firm and stiff; spikes very bristly with long, widely spreading awns, the rachis readily disjointing at the nodes excepting a persistent basal portion; spikelets two at the nodes of the rachis, one of them usually reduced to long awns, the other with two or more grain-producing florets; glumes narrow, divided near the base into two or more hairlike awns, 2–10 cm. long; lemmas smooth or scabrous, tapering into a slender awn 2–10 cm. long.

The spikes of the Squirreltail species may break off near their bases and roll along the ground as tumbleweeds before the wind, gradually breaking up and distributing florets along the way. This genus, of some 4 or 5 species, occurs over most of the western half of the United States.

1. *Sitanion hystrix* (fig. 26) Bottlebrush Squirreltail

Spikes 2–8 cm. long; glumes long, slender, undivided or sometimes divided near the base into 2 long slender awns.

Native. Grows best on dry soils of the mountains but occurs throughout California from sea level to above timberline. Despite its abundance in some areas, Bottlebrush Squirreltail is not a valuable forage grass. The long scabrous awns of the spike are objectionable to grazing animals. It is a poor competitor and only becomes abundant on sandy or rocky soils where other vegetation is scant. In bunchgrass ranges any increase in the density of this species lowers the carrying capacity of that range.

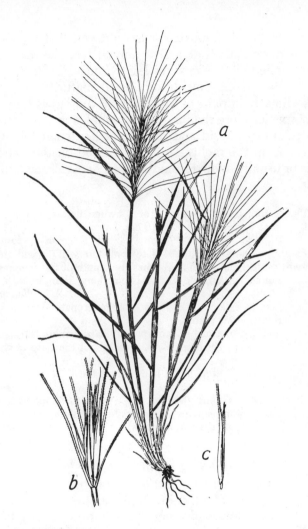

Fig. 26. *Sitanion hystrix*; a, habit of the plant; b, a joint of the spike; c, floret

2. *Sitanion jubatum* (cover, d) Big Squirreltail

Similar to *Sitanion hystrix* differing primarily in the larger and bushier spike, the glumes usually divided near the base into 3 or 4 long slender awns.

This species is widespread throughout California but grows best at lower elevations in the foothills.

16. Genus *Elymus* Wildrye

Mostly perennials, but a few annuals; blades flat or loosely rolled; spikes more or less cylindrical, the rachis, in most species, not disjointing at maturity; spikelets ordinarily in pairs at each node of the rachis but in a few species more than 2; glumes equal, acute or awn-pointed, mostly rigid and sometimes hardened below, persistent on the rachis; lemmas rounded on the back, acute or usually awned from the tip; 12 species in California.

1. Plants annual; spikelets of 1 or 2 awned florets, the awns flattish at least below and 5–10 cm. long...............
   1. *E. caput-medusae*
1a. Plants perennial; spikelets of several florets, if awned the awns scarcely as much as 3 times the length of the lemmas.
   2. Lemma awned, the awn 1–3 cm. long; densely tufted plants............................2. *E. glaucus*
   2a. Lemma essentially awnless, often mucronate; commonly rhizomatous plants.
      3. Lemma scabrous to sparsely hairy; spikes thickish, erect; rhizomes short, close about the base of the plant.
         4. Culms finely hairy about the nodes........
         3. *E. cinereus*
         4a. Culms glabrous..........4. *E. condensatus*
      3a. Lemma smooth; spikes ordinarily slender and becoming curved; plants with long creeping rhizomes
         5. *E. triticoides*

1. *Elymus caput-medusae*
(synonym: *Taeniatherum asperum*)
(fig.27) Medusahead

Annual; culms slender, decumbent to erect, 15–50 cm. tall; blades slender, 3–6 cm. long; spikes 2–6 cm. long (excluding the long awns of the florets and about half as broad; spikelets of 1 to 2 awned florets, the florets readily deciduous at maturity, the awns flat, at least in the lower part and 5–10 cm. long; glumes awlshaped, tapering into an awn 1–2.5 cm long, divergent from the rachis and somewhat incurved.

Introduced from the Mediterranean region. Now naturalized at low to medium elevations in California, it is widespread, occurring on valley plains, foothill flats and slopes, often forming dense patches to the exclusion of most other vegetation. This

Fig. 27. *Elymus caput-medusae*; habit of the plant; an awned floret greatly enlarged; a spike after the florets have fallen away

"crowding out" characteristic makes the grass a serious range weed as it occupies areas where more desirable forage could grow. Livestock ranges with much Medusahead are in poor condition. Good grazing management, mowing, chemical treatments, burning, and reseeding are recommended control measures. Early recognition of infestation is essential to maintain ranges in good condition; once the grass is established in an area it soon spreads by virtue of its aggressiveness and heavy seed crops. It matures somewhat later than the associated annual vegetation; consequently the green heads of Medusahead are conspicuous among the other drying and brownish plants.

The present distribution is mostly in the northern half of the state but with extensions southward in the Sierra Nevada and Coast Ranges into southern California.

2. *Elymus glaucus* (fig. 28) Blue Wildrye

Tufted perennial; culms 50–150 cm. tall, erect to spreading; foliage green or glaucous; blades thin, flat 8–15 mm. wide; spikes 5–20 cm. long, stiffly erect or, in some forms, more or less flexible and seminodding; spikelets 10–12 mm. long; glumes thin, as long as or longer than the florets; lemmas smooth, scabrous or sometimes hairy, tapering into a straight or divergent awn 10–30 mm. long.

Native. Throughout much of California below about 8500 feet; common in the foothills and lower mountain slopes, usually in association with open stands of oaks and conifers. In open areas Blue Wildrye seldom occurs in any kind of a stand. As a forage grass the species is useful early in the season but becomes too "stemmy" later on.

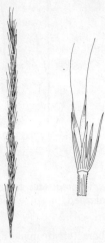

Fig. 28. *Elymus glaucus*; spike and a joint of the spike showing the two spikelets

3. *Elymus cinereus* (pl. 4c) Ashy Wildrye; Basin Wildrye

Plants stout, ordinarily densely tufted, occasionally with a few short rhizomes; culms 60 cm.–2 meters tall, finely short-hairy about the nodes; blades flat, usually less than 15 mm. wide; spikes usually thick and dense with 2–5 spikelets at the nodes, sometimes one of them on a pedicel; spikelets several-flowered; lemmas about 1 cm. long, scabrous to short-hairy, awnless or the tip tapering into a short awn.

Native. Siskiyou, Modoc and Lassen counties south along the east side of the Sierra Nevada; San Bernardino Mountains. Ashy Wildrye is an important forage plant in the sagebrush plant community. It seldom occurs in dense stands, probably because of continual overgrazing.

4. *Elymus condensatus* Giant Wildrye

Robust perennial forming large clumps with short thick rhizomes;

culms stout 2–3 meters tall; blades flat, as much as 3 cm. broad; spikes dense, 15–35 cm. long, mostly compound with branches and hence a dense contracted panicle; spikelets very numerous, several-flowered, the lemmas scabrous to sparsely hairy.

Native. From Alameda and San Mateo counties south along the coast and in the Coast Ranges to San Diego County, east to the San Jacinto and San Bernardino Mountains, thence north to the Tehachapi Mountains. Giant Wildrye seldom occurs in dense stands but is a typical element of the southern California woodlands and brushlands. The large dense panicle is attractive in dried bouquets. As forage, the grass is palatable when young but too coarse by flowering time. Since it occurs in brushlands it is not always accessible to grazing animals.

5. *Elymus triticoides* (pl. 1e, 4d) Creeping Wildrye; Beardless Wildrye

Perennial by extensive creeping rhizomes, commonly gregarious; culms 50–120 cm. tall; foliage green or glaucous; spikes 8–20 cm. long, at first erect but becoming curved at or after maturity; spikelets 12–20 mm. long, the florets rather tardily deciduous; lemmas 6–10 mm. long, smooth, acute or awn-tipped; seed production extremely poor in most areas.

Native. On mostly heavy soils with adequate soil moisture during the growing season: valleys, foothill and mountain flats, and meadows throughout California.

Creeping Wildrye often forms dense stands; through cultivation activities it may be restricted to fence rows or along railroads. In the sagebrush plant association east of the Sierra Nevada crest the grass is frequently valuable as forage, especially in meadow areas that later become dry, where it is grazed by livestock through most of the summer season. The grass resists trampling and recovers well following close grazing. At low elevations it is much coarser than the succulent and tender annuals and is hence seldom utilized by grazing animals. The species is valuable as a soil binder, particularly along levees and river banks.

17. Genus *Agropyron* Wheatgrass

Plants with creeping rhizomes or these wanting and the plants densely tufted; blades flat, often strongly ribbed on the upper surface, or sometimes the blades loosely or even tightly rolled; spike solitary at the apex of the culm, commonly elongate and not at all breaking up at the joints at maturity; spikelets ordinarily large, flattened, solitary at the nodes of the spike, several-flowered, the florets readily falling from the glumes at maturity; glumes firm, several-nerved, pointed at the tips or sometimes blunt; lemmas rounded on the back, usually acute or sometimes awned from the apex, occasionally blunt; 15 native species in California.

The Wheatgrasses grow best in the mountains where they often occur in good stands on dry soils. They are valuable as forage, and their extensive root system holds soil in place. Seed production is ordinarily good in most species and they readily reseed themselves. Several perennial species introduced from Eurasia are proving valuable as forage and in soil stabilization. Three of these are described in the following group of species.

1. Spike with many strongly flattened spikelets crowded on the rachis; glumes awned..............1. *A. desertorum*
1a. Spike with rather widely spaced spikelets in the lower part; glumes acute or obtuse, never awned.
   2. Plants densely tufted.
      3. Lemmas commonly with a curved awn 1–2 cm. long; glumes about half as long as the spikelet..
         2. *A. spicatum*
      3a. Lemmas awnless or at most with an awn much less than 1 cm. long; glumes nearly as long as the spikelet...................3. *A. trachycaulum*
   2a. Plants rhizomatous.
      4. Glumes acute................4. *A. repens*
      4a. Glumes blunt.
         5. Spikelets hairy......5. *A. trichophorum*
         5a. Spikelets glabrous....6. *A. intermedium*

1. *Agropyron desertorum* Desert Wheatgrass

Perennial forming dense tufts; culms slender, 25–80 cm. tall; blades green or bluish-green; spikes 5–9 cm. long, tapering towards the apex; spikelets much flattened, closely placed on the rachis, the lower ones rather widely divergent from the rachis; spikelets of 5–7 florets; tips of the glumes and lemmas with a short awn 2–3 mm. long.

Introduced from Russia in the 1880s. Now naturalized in northeastern California, extending south along the east side of the Sierra Nevada to Inyo County. A valuable forage grass commonly seeded on rangelands in the sagebrush (*Artemisia tridentata*) association.

2. *Agropyron spicatum* Bluebunch Wheatgrass

Perennial forming dense tufts; culms 60–100 cm. tall; blades flat or often tightly rolled, green or blue-green in color; spikes slender, 8–15 cm. long; spikelets 6–8-flowered, rather widely spaced from each other; glumes about half as long as spikelet; lemmas about 1 cm. long commonly with an outwardly curved awn 1–2 cm. long, rarely awnless (var. *inerme*).

Native. Primarily in Lassen, Modoc and Siskiyou counties, thence north to Alaska and east to the Rocky Mountains.

Bluebunch Wheatgrass is a valuable range forage grass. It is of such importance that range condition is based upon its abundance or scarcity. The grass rapidly disappears under successive years of overgrazing and is replaced by inferior species, such as *Sitanion hystrix* or *Bromus tectorum*.

3. *Agropyron trachycaulum* Slender Wheatgrass

Perennial, usually forming dense tufts; culms 50–100 cm. tall; blades green or glaucous, flat to somewhat closely rolled; sheaths commonly smooth, rarely hairy; spikes usually slender, 10–30 cm. long, the spikelets somewhat distant in the lower part; spikelets 3–7-flowered, sometimes as much as 20 mm. long; glumes acute or awn-pointed, nearly as long as the spikelet; lemmas 8–13 mm. long, smooth, acute or short-awned.

Native. On dry to somewhat moist soils commonly at medium to high elevations throughout most of California: higher mountains of southern California, the Sierra Nevada, North Coast Ranges, and Cascade Mountains; thence north to Alaska, east through the United States and Canada to the Atlantic Coast.

A valuable forage grass of summer rangelands where the young growth, stems, and seed heads are usually well grazed by domestic animals. The strong root system of slender wheatgrass resists pulling up by grazing animals and is secondarily valuable in controlling soil erosion, especially on disturbed sites. Overgrazing seriously depletes stands of this grass; consequently some management of herds is necessary in areas where it is plentiful. A good seed bed is necessary when reseeding or a poor stand usually develops.

4. *Agropyron repens* (fig. 29) Quackgrass

Perennial by extensive creeping rhizomes; culms 40–70 cm. tall; blades green or blue-green with short auricles at the base; sheaths commonly short-hairy, the hairs of unequal lengths; spikes narrow, 5–15 cm. long, the spikelets rather closely overlapping; florets 3 to 7 per spikelet; lemmas commonly with pointed tips but in some forms awned.

Quackgrass is a notorious pest in lawns or other ornamental

plantings because of its network of rhizomes. Merely chopping out the stems is not sufficient to control the grass since new shoots soon push their way above the ground from the nodes of the rhizomes. In mountain meadows Quackgrass is sometimes abundant and is useful for pasture; it may be cut for hay.

Fig. 29. *Agropyron repens*; spike and habit of the plant

5. *Agropyron trichophorum*
Stiffhair or Pubescent Wheatgrass

Perennial by creeping rhizomes; culms 60–100 cm. tall; sheaths and blades more or less hairy; spikes stiff, 10–20 cm. long, the spikelets ordinarily hairy; glumes and lemmas blunt.

Introduced from Eurasia. More drought tolerant than *Agropyron intermedium*, *A. trichophorum* has been used successfully in reseeding projects mostly above 3000 feet in the Sierra Nevada and southern California mountains.

6. *Agropyron intermedium*
Intermediate Wheatgrass

Perennial by creeping rhizomes but these ordinarily close about the crown; culms 60–100 cm. tall; blades and sheaths more or less hairy along the margins; spikes stiff, 10–20 cm. long; spikelets lets 3–5-flowered, 10–16 mm. long; glumes glabrous, rather obliquely truncate, 5–7-nerved, shorter than the lowest floret; lemmas 8–10 mm. long, glabrous, awnless.

Introduced from Eurasia. Valuable forage in Lassen and Modoc counties with good stands developing on soils of medium to high fertility. It is also useful in controlling soil erosion.

18. Genus *Secale* Rye
1. *Secale cereale* (pl. 4e, fig. 30) Rye

Stout annual; culms 60–200 cm. tall; blades 6–13 mm. wide, auriculate at the base; spikes slender, 7–15 cm. long, somewhat nodding or at maturity reflexed, rather tardily disjointing at the

Fig. 30. *Secale cereale*; a spikelet with its two florets

nodes in the upper half; spikelets solitary at the rachis nodes, the two florets side-by-side and both maturing plump, oblong grains; glumes narrow, shorter than the lemmas; lemmas stiff-hairy on the keel and margin, tipped by a stout awn about 2 cm. long.

Introduced from the Old World. Now naturalized throughout California; best adapted to light, sandy soils and useful on such sites to control erosion. Rye is sometimes used as a cover crop to be plowed under; it is sometimes grown for forage. Occasionally Rye is weedy in barley and wheat crops.

19. Genus *Avena* Oat

Annuals; ligules membranous, prominent; blades flat, broad; panicles open, spikelets large, drooping in fruit; florets 2 or 3, tough, becoming hardened at maturity, awned from the back of the lemma, awn stout, usually twisted below and once-bent; lemmas bidentate at apex.

1. Lemma glabrous; florets not separating at maturity; awn straight or slightly curved to wanting..........3. *A. sativa*

1a. Lemmas ordinarily long-hairy, at least about the callus; florets readily separating at maturity; awn twisted below and bent above.
 2. Lemma teeth acute; spikelets ordinarily 3-flowered..
....1. A. *fatua*
 2a. Lemma teeth attenuate as bristles; spikelets ordinarily 2-flowered..........................2. A. *barbata*

Fig. 31. *Avena barbata*; panicle

1. *Avena fatua* (fig. 32a) Wild Oat

Culms 25–100 cm. tall; foliage sparsely hairy; spikelets on curved, hairlike pedicels; glumes thin, several-nerved, 6–8 mm. broad (at about the middle); florets usually 3, or in robust specimens 4, rarely only 2, readily separating at maturity; lemma apex with 2 short, acute teeth about 1 mm. long; base of lemma with long, brownish to whitish hairs.

Introduced from the Old World. Now naturalized at low elevations throughout much of California; grows best on rich soils of the valleys where it is often robust. While on thin or sandy soils it is depauperate and produces only a few spikelets per plant. It is a common weed of towns, along roads, fencerows, in old fields, etc. On the valley and foothill rangelands it is valuable as winter and spring forage. An abundance of Wild Oats indicates good range condition while a scarcity of the grass suggests a decline in desirable forage species. The panicles are decorative and frequently used in dry bouquets.

2. *Avena barbata* (figs. 31, 32b) Slender Oat

Similar to A. *fatua* but most forms have glabrous foliage; glumes

a. *Agrostis diegoensis*

b. *Stipa speciosa*

c. *Stipa speciosa* habitat

d. *Koeleria cristata*, old bunchgrass

e. *Elymus triticoides*

f. *Bromus rubens*

g. *Stipa pulchra* grassland

h. *Stipa cernua*

PLATE 1

a. Coastal grassland

b. *Festuca arundinacea*

c. *Festuca idahoensis* (center) and *Stipa elmeri*

d. *Lolium multiflorum*

e. *Poa pratensis*

f. *Poa scabrella*

g. *Poa nevadensis*

h. *Briza maxima*

PLATE 2

a. *Melica californica*

b. *Bromus willdenowii*

c. *Bromus tectorum*

d. *Bromus diandrus*

e. *Bromus rubens*

f. *Triticum aestivum*

g. *Aegilops triuncialis*

h. *Hordeum leporinum*

PLATE 3

a. *Hordeum vulgare*

b. *Hordeum geniculatum*

c. *Elymus cinereus*

d. *Elymus triticoides*

e. *Secale cereale*

f. *Avena sativa*

g. *Holcus lanatus*

h. *Deschampsia caespitosa*

PLATE 4

a. *Anthoxanthum odoratum*

b. *Agrostis avenacea*

c. *Polypogon monspeliensis*

d. *Phalaris tuberosa* var. *stenoptera*

e. *Phalaris minor*

f. *Oryzopsis miliacea*

g. *Oryzopsis hymenoides*

h. *Ampelodesmos mauritanicus*

PLATE 5

a. *Phyllostachys bambusoides* var. *aurea*

b. *Oryza sativa*

c. *Ehrharta calycina*

d. *Arundo donax*

e. *Phragmites australis*

f. *Cortaderia selloana*

g. *Muhlenbergia rigens* habitat

h. *Leptochloa fascicularis*

PLATE 6

a. *Leptochloa uninervia*

b. *Distichlis spicata* vegetative

c. *Distichlis spicata* flowering

d. *Neostapfia colusana* habitat

e. *Panicum capillare* habitat

f. *Panicum hillmanii*

g. *Digitaria sanguinalis* vegetative

h. *Paspalum distichum* habitat

PLATE 7

a. *Pennisetum villosum*

b. *Pennisetum setaceum*

c. *Setaria glauca*

d. *Andropogon virginicus*

e. *Sorghum halepense* in orchard

f. *Sorghum sudanense*

g. *Sorghum bicolor*

h. *Zea mays*

PLATE 8

not as broad; florets 2 in number; lemma apex tipped with 2 slender hyaline bristles 3-4 mm. long.

Introduced from Europe. Now naturalized throughout most of California at low elevations; widely distributed over foothill slopes and quite tolerant of a variety of soils. Slender Oat is one of the preferred forage grasses on California's winter annual range. The seedlings are strong and vigorous, producing much succulent feed for grazing animals during spring. As A. *barbata* approaches maturity and with the eventual shedding of the seed, it rapidly loses protein and its appeal to grazing animals. The panicles are decorative in dried flower arrangements.

3. *Avena sativa* (pl. 4f) Oat

Similar to *Avena fatua,* differing mostly in the 2-flowered spikelets and the florets not breaking apart nor separating from the glumes; lemmas ordinarily hairless, awnless or with a straight or slightly curved awn.

Cultivated as a crop and occasionally sporadic along roads, in towns, edges of fields. It may occur anywhere in the state from low to high elevations.

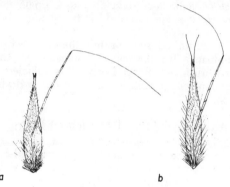

Fig. 32. Florets of *Avena;* a, *Avena fatua;* b, *Avena barbata*

20. Genus *Aira* Hairgrass
1. *Aira caryophyllea* (fig. 33) Silver Hairgrass

Delicate annual 5-25 cm. tall; blades short, few in number; ligules membranous, prominent; panicles at maturity open, spikelets whitish to silvery, clustered towards the ends of the slender branches; glumes about 3 mm. long, enclosing two slender, rough, awned florets; lemmas tipped with 2 slender bristles.

Introduced from Europe. Naturalized over most of California at low to medium elevations, on continually overgrazed annual

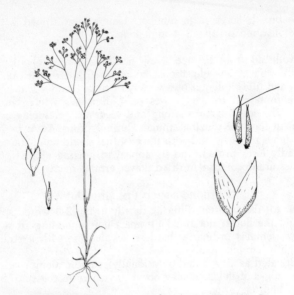

Fig. 33. *Aira caryophyllea*; spikelet and a single floret; habit
Fig. 34. *Aira elegans*; spikelet and the two florets

ranges or on disturbed soils or infertile sites, where it soon assumes dominance. It is particularly common along trails, roads, or in cleared brushland. Silver Hairgrass is valueless as forage, and an abundance of it indicates overgrazing or decline in fertility or both.

2. *Aira elegans* (fig. 34) Elegant Hairgrass
Annual; similar in most respects to *A. caryophyllea*, differing primarily in the more diffuse panicle, spikelets widely scattered at the ends of the branches, and the lowest of the two florets awnless. The glumes are 2.5 mm. long.

Introduced from Europe. Now naturalized mostly at low elevations in northern California. It occupies similar sites as Silver Hairgrass and is valueless as forage.

21. Genus *Holcus* Velvetgrass
1. *Holcus lanatus* (pl. 4g; fig. 35) Common Velvetgrass
Usually tufted perennial with some forms developing short rhizomes; foliage densely and softly pubescent; culms 25–75 cm. tall; panicles contracted, variable as to density, often reddish-purple to pinkish or sometimes pale green to whitish; spikelets readily falling entire from their pedicels; glumes unequal, enclos-

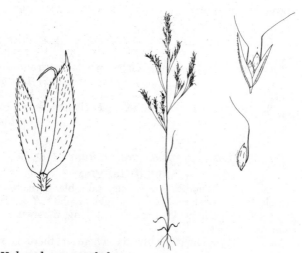

Fig. 35. *Holcus lanatus* spikelet
Fig. 36. *Deschampsia danthonioides*; habit, spikelet and a single floret

ing two unlike florets; lower floret grain-bearing, smooth, awnless, the pedicelled one smaller with a hooked or wavy awn about 3 mm. long.

Introduced from Europe. Now naturalized throughout much of California at low elevations and ascending into the mountains at medium or occasionally high elevations. Common Velvetgrass grows best on moist soils or at least those soils with subsurface moisture during the growing season. It is a poor forage grass because of its exceptionally light "body." Early growth is readily cropped by grazing animals but upon elongation of the culm and flowering it loses its appeal to these animals. Dense stands of Velvetgrass often occur in meadows, but it yields poor quality hay. The grass gradually increases in density and subsequently crowds out more desirable forage.

22. Genus *Deschampsia* Hairgrass

Mostly tufted perennials, a few annuals; panicles open or contracted, the spikelets more or less shiny, mostly 2-flowered; glumes slightly unequal, acute; lemma shiny, hairy about the callus, the apex obtuse, with 2–5 short teeth, awned from the back, the awn twisted below and once-bent at about the middle or sometimes straight; rachilla conspicuously long-hairy, prolonged above the uppermost floret. Five native species in California.

1. Annual..........................1. *D. danthonioides*
1a. Perennial.
   2. Basal leaves firm; panicles open to dense and thickish..
      2. *D. caespitosa*
      3. *D. holciformis*
   2a. Basal leaves filiform, soft; panicle narrow and elongate.
      4. *D. elongata*

1. *Deschampsia danthonioides* (fig. 36)
Annual Hairgrass

Annual; culms slender, 10–40 cm. tall; blades narrow, short; panicle large in relation to the height of plant, 7–25 cm. long; glumes 4–8 mm. long; lemma 2–3 mm. long, the awn protruding from the glumes.

Native. Throughout California wherever there is sufficient moisture present during the growing season at all elevations. This delicate grass produces little foliage hence is poor as range forage but the open panicles are sometimes colorful especially when the spikelets are dark reddish-purple. In the Sacramento Valley, Annual Hairgrass is conspicuous in vernal meadows of low areas. It is frequent about the edges of rain pools sometimes forming a purplish ring about the pool in the spring.

2. *Deschampsia caespitosa* (pl. 4h)
Tufted Hairgrass

Densely tufted perennial; culms erect to widely spreading 30–120 cm. tall; panicles open to loosely contracted, 10–25 cm. long; glumes 4–5 mm. long; lemmas 3–4 mm. long; awn hairlike, straight or nearly so, 2–4 mm. long, arising from near the base.

Native. Higher mountains of southern California north in the Sierra Nevada and Cascades; Coast Ranges from Santa Barbara County north; thence to Canada, Alaska, east to the Rocky Mountains and Atlantic Coast. Tufted Hairgrass is a principal constituent of mountain meadows, but it also occurs at lower elevations and along the coast in springy areas, seeps, marshes, and bogs. It is one of the most important range species in the western United States since all classes of livestock forage on the plant; it is widely distributed, frequently in dense stands, it can withstand rather close grazing and recovers well. It is also valuable in control of soil erosion.

Fig. 37. *Deschampsia holciformis*; a, panicle; b, glumes; c, florets

3. ***Deschampsia holciformis*** (fig. 37) Pacific Hairgrass
Similar to *Deschampsia caespitosa* and perhaps not specifically distinct but typically has a dense, narrow panicle.
Native. Along the coast from Santa Barbara County north to Del Norte County; thence north to British Columbia.

4. ***Deschampsia elongata*** (fig. 38) Slender Hairgrass
Short-lived perennial; culms slender, 30–80 cm. tall; blades hairlike and mostly in a basal tuft; panicle narrow, 10–40 cm. long with some long branches below but these branches, as well as the upper ones, closely appressed to the rachis; spikelets 4–6 mm. long; lemma 2–3 mm. long; awn hairlike, straight, exceeding the lemma.
Native. On moist soils from lower to higher elevations throughout much of California, thence north to Alaska, east to the Rocky Mountains, south through Mexico and thence to South America. Slender Hairgrass occurs in a variety of habitats but always where there is sufficient moisture during the growing season, and seldom in dense stands. It is a poor competitor among other plants.

23. Genus *Koeleria* Koeleria

1. ***Koeleria cristata*** (pl. 1d; fig. 39) Prairie Junegrass
Tufted perennial or some forms with short rhizomes; culms stiff, erect to slightly curved, 15–60 cm. tall; foliage green to gray-green, variously hairy; panicle shiny, dense, spikelike, often interrupted in the lower part, 3–10 cm. long, silvery to pale green or purplish; spikelets strongly flattened, 2–3 flowered, lemmas awnless.

Fig. 38. *Deschampsia elongata* habit

Native. On dry soils throughout California ranging from the oak woodlands of the foothills to the high mountain peaks. Koeleria provides good forage for livestock and grows best in the mountains. The usually poor seed set has greatly reduced the stands of Prairie Junegrass, particularly on overgrazed rangelands where recovery both by seedlings and parent plants is essential to maintain good range condition.

24. Genus *Trisetum* Trisetum

Mostly tufted perennials; blades flat; panicle shiny,

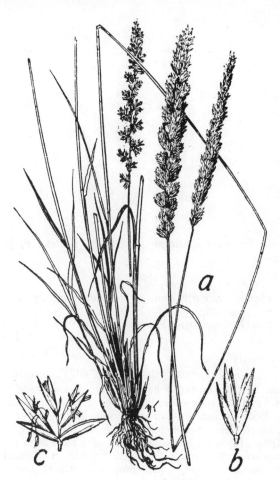

Fig. 39. *Koeleria cristata*; a, habit; b, spikelet; c, flowering spikelet

open, contracted or spikelike; spikelets flattened, 2-, sometimes 3–4-flowered; rachilla densely hairy, prolonged beyond the uppermost floret; glumes commonly unequal, the longer one equaling or exceeding the florets; lemma hairy above the callus, 2-toothed at the apex, the teeth attenuated into awns, the midrib prolonged into a straight or bent awn arising in the upper third of the lemma.

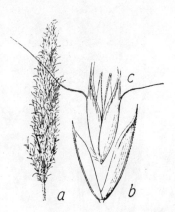

Fig. 40. *Trisetum spicatum*; a, spikelike panicle; b, glumes; c, florets

1. Panicle ordinarily loose to contracted, mostly exceeding 10 cm. in length.........................1. *T. canescens*
1a. Panicle dense, spikelike, usually less than 10 cm. long....
2. *T. spicatum*

1. *Trisetum canescens* Tall Trisetum

Culms 50–100 cm. tall; foliage variously hairy to scabrous only; blades flat, 2–7 mm. wide; in the widest sense the panicle varies from loose to contracted or rarely spikelike, 10–20 cm. long, 2–3-flowered; lemmas 5–6 mm. long.

Native. Common in wooded areas, high mountains of southern California north through the Sierra Nevada and Cascades; from Santa Barbara County north through the Coast Ranges.

2. *Trisetum spicatum* (fig. 40) Spike Trisetum

Culms 10–40 cm. tall; leaves mostly basal, variously hairy to glabrous; panicle 5–10 cm. long, pale green, bronze-colored or purplish, dense spikelike; spikelets 4–6 mm. long.

Native. Ridges, slopes and flats of the mountains at elevations above 6000 feet; high mountain ranges of southern California, Sierra Nevada, Cascades, North Coast Ranges; thence north to Alaska and east.

Spike Trisetum is a valuable forage grass of the summer rangelands, as the succulent foliage and soft seed heads are utilized by grazing animals throughout the season.

25. Genus *Anthoxanthum* Vernalgrass
1. *Anthoxanthum odoratum* (pl. 5a; fig. 41)
Sweet Vernalgrass

Short-lived perennial emitting a sweetish odor especially when cut fresh; culms slender, 30–60 cm. tall; foliage hairy in some forms but more often nearly hairless; blades flat; panicle dense to somewhat loose below, cylindric, tapering toward apex, 3–8 cm. long; glumes unequal, the uppermost 8–10 mm. long, the lower one about half as long; florets 3, the lower 2 sterile, dark-brown, hairy, awned and mostly enclosing a smooth, awnless, grain-producing one. At maturity the 3 florets readily fall as a unit from between the glumes.

Introduced from Europe. Naturalized over much of northern California, most common along the northern coast and meadows of the interior foothills and mountains up to about 6000 feet elevation. The grass often occurs in fairly dense stands but is low quality forage either as pasture or hay.

Fig. 41. *Anthoxanthum odoratum* spikelet showing the glumes right and the awned sterile florets left

26. Genus *Agrostis* Bentgrass

Perennials or sometimes annuals; blades commonly flat; panicles open to loosely contracted, rarely almost spikelike; spikelets 1-flowered, the floret ordinarily deciduous from between the glumes; lemmas mostly 3-nerved, awned from the back or awnless; paleas mostly poorly developed as a thin scale or sometimes this wanting; callus hairs poorly developed.

A large genus of about 150 species distributed throughout the cool regions of the world. In California about 25 species occur in widely diverse habitats, plant associations, and elevational range. They are valuable as range forage in the mountains most species preferring moist habitats. Several species are useful as turfgrass and others as forage in irrigated pastures.

1. Rhizomatous.................................1. A. *alba*
1a. Tufted.
   2. Panicle open, diffuse.
      3. Low elevations; lemma hairy....2. A. *avenacea*
      3a. Medium to high elevations; lemma without hairs.
                                    3. A. *scabra*
   2a. Panicle dense or loosely contracted......4. A. *exarata*

1. *Agrostis alba* (fig. 42) Redtop

Perennial, commonly forming dense tufts but with extensive development of short rhizomes about the tuft; culms 35–80 cm. tall; blades flat, 5–10 mm. wide; panicles somewhat pyramidal in outline to oblong; glumes 2–3 mm. long; lemmas 2/3 to 3/4 length of glumes.

Introduced from Europe first as a pasture grass. Now naturalized and widespread over most of the state from coastal to high elevations in the interior mountains. Cattle and horses are especially fond of it, but it fairs better in growth and as hay forage when it is mixed with other grasses and legumes. At medium to high elevations in the mountains it remains green through most of the season and recovers well from close grazing. Redtop is valuable in retaining soil on roadbanks and along ditches.

2. *Agrostis avenacea* (pl. 5b; fig. 43a) Pacific Bentgrass

Short-lived perennial; culms 30–60 cm. tall; culm leaves with a membranous ligule 3–8 mm. long; panicles diffuse, 15–30 cm. long, soon breaking away from the culm and behaving as a tumbleweed; glumes 3–4 mm. long, both of them 1-nerved; lemmas hairy, 1.5–2 mm. long, a fine awn arising from about middle of back, once-bent, and exceeding glumes.

Introduced from Australia. Naturalized mostly in the Central Valley but extending into the surrounding foothills, Delta region,

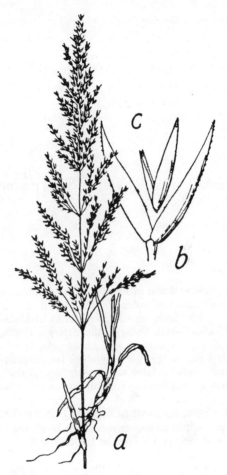

Fig. 42. *Agrostis alba*; a, habit; b, spikelet; c, floret placed above the glumes, palea at the left and lemma to the right

and around the San Francisco Bay; occurs abundantly in old rice fields or pastures and marshlands. By late spring it has widely dispersed its airy, wind-transported panicles, which pile up in ditches and along fences conspicuously.

3. *Agrostis scabra* (fig. 43b) Ticklegrass

Tufted perennial; culms 20–60 cm. tall; basal foliage fine, slender, those of culms longer, wider; panicles ordinarily diffuse, 15–25 cm. long, branches elongate, slender, very scabrous, bearing small

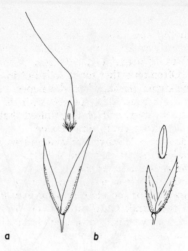

Fig. 43. *Agrostis* spikelets; a, *Agrostis avenacea*, b, *Agrostis scabra*

spikelets toward ends; glumes unequal; lemmas 1.5 to 1.7 mm. long, usually awnless but occasionally with a straight awn.

Native. Moist soils, mostly medium to high elevations, southern California, north through the Sierra Nevada, North Coast Ranges; thence to Canada and Alaska; also throughout much of the United States. It is good livestock forage early in the season, but animals avoid it later after emergence of the panicle, which is airy and slender but has strongly scabrous branches that tickle.

4. *Agrostis exarata* (fig. 44) Spike Bentgrass
Tufted perennial, quite variable as to form, some plants of low to medium height, others robust and as much as 120 cm. tall, but

Fig. 44. *Agrostis exarata* spikelet showing the glumes and the single floret

Fig. 45. *Calamagrostis canadensis* spikelet

ordinarily the culms are 30–60 cm. tall; blades flat; panicles variable, nearly always dense but quite variable in length and thickness, ordinarily 8–15 cm. long; glumes 2–4 mm. long, rather

sharply acute to awn-tipped and more or less scabrous; lemmas
1.7–2.0 mm. long, awned or commonly the awn rudimentary
requiring strong magnification to see.

Native. Often on rather moist soils but in various habitats and
plant associations, perhaps best developed in the mountains and
along the north coast. The species, with its many forms, occurs
through the western half of the United States; thence north to
Canada and Alaska; south into Mexico.

Spike Bentgrass is particularly abundant along streams, in or
about meadows, moist slopes or moist clearings in the coniferous
forest. The green foliage and soft seed heads are utilized by
grazing animals in the mountains during most of the summer,
though excessive or too close grazing may markedly reduce the
density of the stand. On disturbed soils the grass soon forms a
good stand but over successive years gradually wanes because of
competition from other plants.

27. Genus *Calamagrostis* Reedgrass

1. Panicle open to somewhat contracted and, if the latter, then the callus hairs nearly as long as the lemma............
   1. *C. canadensis*
1a. Panicle loosely contracted; callus hairs less than 1/2 as long as the lemma........................2. *C. nutkaensis*

1. *Calamagrostis canadensis* (fig. 45)
Bluejoint Reedgrass

Perennial by rhizomes; culms 40–100 cm. tall; blades flat, scabrous, 4–8 mm. wide; panicles 10–25 cm. long, commonly nodding and variably open to somewhat contracted; glumes purplish 3–4 mm. long; floret 1, the lemma with a fine hairlike awn arising from the back below the middle and equaling the apex; callus hairs copious, nearly as long as the lemma; rachilla with long silky hairs. At maturity the light florets are carried away by the wind, the long hairs acting as the "parachutes."

Native. At medium to high elevations in the Sierra Nevada, North Coast Ranges and Cascades, thence north and east beyond the state. Bluejoint Reedgrass occurs on moist soils of meadows, slopes, or open woods above 5000 feet elevation. During vegetative growth the grass provides succulent forage. The extensive underground rhizomes allow for rapid recovery following close grazing.

2. *Calamagrostis nutkaensis* Pacific Reedgrass

Tufted perennial with some short rhizomes about the base and forming tough clumps; culms 1–1.5 meters tall; blades scabrous, flat or rolled, elongate; ligule 4–6 mm. long; panicles mostly loosely contracted, purplish to brownish, eventually becoming straw-colored, 15-30 cm. long; glumes 5-7 mm. long; lemma 4-5 mm. long, the awn hairlike, arising near the base; callus hairs less than half as long as the lemma.

Native. Mostly on the coastal slopes, meadows, and bogs but occasionally inland along the outer North Coast Ranges to about 5000 feet elevation. Del Norte County south along the coast to about Point Conception in Santa Barbara County; north to Alaska.

The grass is valuable in controlling soil erosion on steep bluffs and slopes above the sea. As livestock forage it is usually too coarse, although the young growth is quite palatable.

28. Genus *Polypogon* Polypogon

Introduced annuals or short-lived perennials; blades flat, scabrous; panicle dense, commonly spikelike, sometimes lobed, soft bristly; spikelets 1-flowered and readily deciduous from the pedicel as a unit; glumes with a fine, straight or flexuous awn; lemma hyaline with a fine awn arising below the apex or awnless; grain ovoid.

1. Annuals; glumes ciliate about the apex.
    2. Lemma awned.................1. *P. monspeliensis*
    2a. Lemma awnless...................2. *P. maritimus*
1a. Perennial; glumes not at all ciliate about the apex........
    3. *P. australis*

1. *Polypogon monspeliensis* (pl. 5c; fig. 46)
Rabbitfootgrass

Annual; culms 10-50, rarely to 80, cm. tall; sheaths loose on the culm, blades scabrous 3-8 mm. wide; ligule 4-6 mm. long; panicles very bristly and usually silky but sometimes pale- to yellowish-green, rarely purplish, 2.5-7 cm. long (in occasional robust specimens up to 15 cm. long), mostly emergent from the upper sheath but sometimes included at the base; glumes about 2 mm. long; lemmas smooth, shiny, about half as long as glumes.

Introduced from Europe. Now naturalized on moist or wet soils throughout California at low elevations; common around seeps, springs, stream and river beds, ditches, meadows, roadsides, and old pastures. It grows equally well on alkaline, saline or neutral soils. Rabbitfootgrass is primarily a weed, but on the alkaline plains of the Central Valley livestock occasionally graze the plant. The panicle is distinctive and readily identifies the grass.

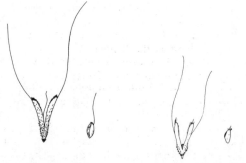

Fig. 46. *Polypogon monspeliensis*; spikelet and the small, awned floret
Fig. 47. *Polypogon maritimus*; spikelet and floret

2. *Polypogon maritimus* (fig. 47)
Mediterranean Polypogon

Annual; similar in some respects to *Polypogon monspeliensis* but differing consistently by: more slender blades (1–3 mm. wide), longer lobed (above the insertion of the awn) glumes, and awnless lemmas.

Introduced. Now naturalized, mainly in northern California on moist soils at low elevations, and probably at scattered locations in southern California.

3. *Polypogon australis* Chilean Polypogon

Perennial; culms 50–100 cm. tall; blades 4–10 mm. wide and up to 18 cm. long; ligule truncate 1–2 mm. long; panicles very bristly even to being "bushy," strongly lobed, interrupted below, greenish or more often purplish, 5–15 cm. long; glumes 2–2.5 mm. long, often divaricate exposing the plump floret, the glume awn flexuous, 5–7 mm. long; lemmas about 1 mm. long, shiny, awned from the back, the awn flexuous and about 4 mm. long.

Introduced. Now naturalized on moist soils throughout California at low elevations; common along streams and ditches especially at the edge of water.

29. Genus *Gastridium* Nitgrass
1. *Gastridium ventricosum* (fig. 48) Nitgrass

Annual; culms 10–40 cm. tall, rather wiry; blades flat, relatively few in number; panicles spikelike, dense, pale, shiny, cylindrical, tapering at the apex; glumes persistent, unequal, 3–5 mm. long, rather swollen below to accommodate the single, globular, hairy

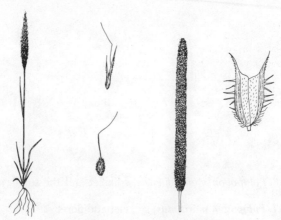

Fig. 48. *Gastridium ventricosum*; habit of the plant, a spikelet and the awned floret

Fig. 49. *Phleum pratense*; panicle and a spikelet

floret; awn hairlike, 5 mm. long.

Introduced from Europe. Now naturalized on dry foothills and open plains at low elevations throughout California. An abundance of Nitgrass on foothill ranges indicates overuse by grazing animals, since the better forage species have been grazed down and afford much less competition for the development of Nitgrass. On disturbed soils it is abundant at first but in succeeding years gradually decreases in density. Nitgrass commonly occurs about stands of brush and when the brush is cleared away the grass develops robust plants for the species. The low stature, wiriness, and general lack of nutritious foliage rates it poor to nil as forage for livestock. Nitgrass is easily recognized and is quite durable up to and even past the first rains of fall.

30. Genus *Phleum* Timothy

Tufted perennials with flat blades and cylindric, spike-like panicles; spikelets strongly flattened; glumes equal, truncate, the midnerve prolonged as a stout awn 1–2 mm. long, midnerve (keel) also coarsely hairy below the blunt apex; lemma thin, ovoid, much shorter than the glumes.

1. Panicle 5 cm. or more long.................1. *P. pratense*
1a. Panicle less than 3 cm. long..............2. *P. alpinum*

1. *Phleum pratense* (fig. 49) Timothy

Forming dense clumps; culms swollen at the base, 50–100 cm. tall; blades 7–20 cm. long; panicle 5–10 cm. long, 5–8 mm. wide; glumes 3.5 mm. long.

Introduced from Europe. Now naturalized and best adapted to cool mountain habitats, mostly in or at the edge of meadows throughout most of California above about 3000 feet in elevation. As a range plant good stands are sometimes obtained by seeding, but they are difficult to maintain because of excessive grazing of the grass. Timothy is a valuable hay and pasture grass in the cooler parts of the United States.

2. *Phleum alpinum* Alpine Timothy

Forming loose tufts; culms 20–40 cm. tall; panicle 1.5–2.5 cm. long, 7–13 mm. wide; glumes about 5 mm. long, awns about 2 mm. long.

Native. In bogs, meadows, along streambanks, mostly in the high mountain ranges throughout California; east to the Rocky Mountains, north to Canada; Europe, Asia. It occurs also along the immediate coast of California from about San Francisco Bay north. It is a valuable forage grass in the mountains but stands are rather sparse.

31. Genus *Phalaris* Canarygrass

Annuals or perennials; blades flat; ligule membranous, well developed; panicle usually dense, spikelike, oblong, sometimes lobed in robust specimens; spikelets strongly flattened; glumes equal, tough, boat-shaped, 3-nerved, sometimes winged on the keel below the apex; fertile floret more or less pear-shaped, appressed-hairy at least in the lower half; sterile florets commonly 2, rarely 1, closely appressed to the fertile one. In *Phalaris paradoxa* the sterile florets are wanting.

1. Perennial with somewhat swollen culm bases and sometimes short rhizomes as well....................1. *P. tuberosa* var. *stenoptera*
1a. Annual.
   2. Clusters of sterile spikelets surround a fertile spikelet and the whole cluster of both kinds of spikelets falls entire from the panicle............2. *P. paradoxa*
   2a. Spikelets all grain-producing and not falling entire from the panicle........................3. *P. minor*

1. *Phalaris tuberosa* var. *stenoptera* (pl. 50; fig. 50b)
Hardinggrass

Perennial; forming large clumps with some very short rhizomes about the base; culms 60–150 cm. tall, somewhat swollen at the base; panicle 5–15 cm. long, dense, spikelike, sometimes lobed in larger panicles; glumes 5–6 mm. long, narrowly winged on the keel; fertile floret 3.5–4 mm. long, sterile florets commonly 2, one of them minute or obsolete.

Introduced through Australia and probably native of the Mediterranean region. Now more or less naturalized at low elevations in California. Hardinggrass is one of the most valuable forage grasses seeded on the valley and foothill rangelands in California. It is useful in seeding prepared land in the oak woodland of the foothills and cleared brushlands. Hardinggrass is slow in developing but after several years forms sizeable bunches and reseeds itself fairly well but not uniformly over any given range.

The value of seeding this perennial on typically annual forage rangeland lies in its prolonged period of green forage beyond the time annuals have completely dried. Established plants green up before the fall rains and provide much succulent forage during the winter when the annuals are yet small.

2. *Phalaris paradoxa* (fig. 50c) Hood Canarygrass

Annual; culms 30–60 cm. tall; panicle dense, spikelike, more or less club-shaped, 2–6 cm. long, very often partially enclosed at the base by the uppermost leaf sheath; spikelets in the upper 2/3 of the panicle falling away at maturity in clusters of 6 or 7 and all but 1 of the cluster are sterile spikelets; fertile floret 3 mm. long, the sterile florets obsolete.

Introduced from Europe. Now naturalized as a common weed of richer valley soils, occasional along roads or about cities and towns.

3. *Phalaris minor* (pl. 5e; fig. 50a)
Littleseed Canarygrass; Mediterranean Canarygrass

Annual; culms 30–60 cm. tall, rarely as robust plants to 1 meter tall; panicles dense, spikelike, oblong, 2–5 cm. long; glumes 4–6 mm. long, the thin wings on the keel more or less irregularly notched; fertile floret nearly 3 mm. long, 1 sterile floret about 1 mm. long.

Introduced from Europe. Now naturalized as a common weed on rich valley soils, occasional along roads or in and about towns and cities.

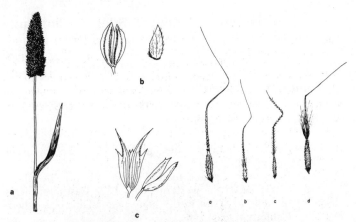

Fig. 50. a, *Phalaris minor*; b, *Phalaris tuberosa* var. *stenoptera* spikelet and florets; c, *Phalaris paradoxa* fertile and sterile spikelets

Fig. 51. Florets of *Stipa*; a, *Stipa pulchra*; b, *Stipa californica*; c, *Stipa occidentalis*; d, *Stipa speciosa*

32. Genus *Stipa* Stipa

Tufted perennials (California species); culms erect with narrow, usually involute basal blades; panicle often narrow-contracted, sometimes more or less open or at least with some spreading branches below; floret 1, hairy, readily deciduous from the spikelet; glumes thin, commonly acuminate, longer than the floret; lemmas leathery, enclosing the much shorter palea, becoming hardened in fruit; lemma awn prominent, stout, tightly twisted below and once or twice bent above, scabrous or hairy; callus usually sharp-pointed, densely short-hairy above the glabrous point.

Twenty species of Stipa occur in California, most of them in the mountains above 3500 feet. They are well adapted to droughty soils and are generally widespread but not usually in very dense stands. Stipas green up early in the season and provide much early forage for grazing animals. They need some protection from grazing at the flowering period to ensure formation of the seed and also to allow for storage of food reserves in the crown. Stands of Stipa are ordinarily maintained by the abundant seed production in nongrazed or sometimes lightly grazed areas. The sharp-pointed florets are augered into the soil by the twisting and untwisting of the long awns. Stipas soon invade disturbed soils and because of their strong root system are valuable in erosion control.

1. Panicle rather loose to open, at least the lower branches long and spreading to drooping in fruit.
    2. Plants of valley and foothill slopes mostly below about 3500 feet on the western slope of the Cascades, Sierra Nevada, and also in the higher mountains of southern California.
        3. Lemma 4–6 mm. long, awn slender, less than 5 cm. long..........................1. S. *lepida*
        3a. Lemma over 7 mm. long, awn stoutish, over 5 cm. long.
            4. Terminal segment of the awn straight; basal blades flat...................2. S. *pulchra*
            4a. Terminal segment of the awn slender and flexuous; basal blades slender...3. S. *cernua*
    2a. Plants of semi-arid mountain slopes or flats above about 4500 feet; eastern slope of the Cascades and Sierra Nevada.............................4. S. *comata*
1a. Panicle narrow-contracted, the branches always closely appressed to the main axis.
            5. Awn once-bent, densely long-hairy on the lower segment; plants of arid regions.
                5. S. *speciosa*
            5a. Awn twice-bent, if hairy then rather closely so and at least on 2 segments of the awn; plants of dry mountain slopes and flats.
6. Awn hairy on all three segments of the awn..............
    6. S. *occidentalis*
6a. Awn hairy on the two lower segments of the awn or all three segments are scabrous.
    7. Blades and sheaths finely and densely short-pubescent; awn prominently and densely hairy on the lower two segments............................7. S. *elmeri*
    7a. Blades and sheaths ordinarily glabrous (finely pubescent around the collar in *Stipa californica*); awn scabrous to loosely hairy.
        8. Glumes narrow and long-tapering, the lower one

3-nerved; callus sharply pointed; plants of medium to high elevations.
- 9. Lemma hairs conspicuously longer toward the apex of the lemma than those below; awn more or less loosely pubescent on the lower two segments..............8. *S. californica*
- 9a. Lemma hairs all about the same length; awn scabrous only.............9. *S. columbiana*
- 8a. Glumes rather broad at the base and abruptly tapered above, the lower 5-nerved; callus blunt; plants of lower to medium elevations..........
10. *S. lemmoni*

1. *Stipa lepida* (fig. 53b) Foothill Stipa

Densely tufted; culms finely pubescent below the nodes, 40-90 cm. tall; sheaths somewhat hairy about the throat; panicles loose to open or at least with some long lower branches and 10-40 cm. in length; glumes 6-10 mm. long; lemma 4-6 mm. long, sparingly hairy in the lower half and almost glabrous below the apex, the apex with a ciliate crown.

Native. Common in brushlands from San Diego County north in the Coast Ranges to Humboldt County; Sierra Nevada foothills from about Madera County north to Shasta County.

2. *Stipa pulchra* (cover-c; fig. 51a) Purple Stipa

Densely tufted; culms 40-90 cm. tall; foliage glabrous to finely hairy, basal blades flat, 2.4-6 mm. wide; panicles loose to open, the lower branches relatively long, spreading and drooping in fruit; glumes 14-20 mm. long; lemma 7.5-13 mm. long, sparingly appressed-hairy, often the hairs disposed in lines towards the short-ciliate crown; awn twice-bent, 5-8 cm. long, scabrous to short-hairy, stoutish in the lowermost segment.

Native. Central Valley and surrounding foothills; central coast and south to San Diego County. One of the major species of the California grassland.

3. *Stipa cernua* (pl. 1h) Nodding Stipa

Similar in most respects to *Stipa pulchra*, differing mostly in the finer blades and flexuous terminal segment of the awn. The lowest segment of the awn is very slender.

Native. About the same distribution as *Stipa pulchra*.

4. *Stipa comata* (fig. 52) Needle-and-Thread

Densely tufted; culms 30-60 cm. tall; ligule 3-5 mm. long; panicles often included at the base and sometimes almost completely included; glumes 2-3 cm. long; lemma 9-13 mm. long, finely appressed-hairy; awn mostly flexuous, 10–15 cm. long.

Native. Western edge of the Mojave Desert north along the eastern slope of the Sierra Nevada and Cascades; thence north and east.

The var. *intermedia* differs from the typical in the ordinarily well-exserted panicle, more distinctly bent awn with a straight third segment. The glumes and lemmas are a little longer than the typical.

Native. Within the range of the species.

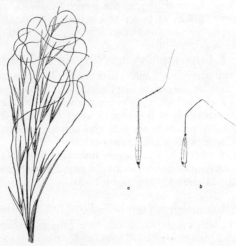

Fig. 52. *Stipa comata* panicle
Fig. 53. *Stipa* florets; a, *Stipa elmeri*; b, *Stipa lepida*

5. *Stipa speciosa* (pl. 1c, 1d; fig. 51a) Desert Stipa

Densely tufted; culms 30–60 cm. tall; blades slender, at length becoming tough; panicle narrow-contracted, 10–20 cm. long, often partially included in the upper leaf sheath; glumes 1.5–2 cm. long; lemmas 7–9 mm. long, evenly short-hairy; awn once-bent, densely long-hairy on the lower segment, the hairs 4–7 mm. long.

Native. Arid and semi-arid mountain slopes: mountains of southern California and east; north along the east side of the Sierra Nevada to Alpine County; Tehachapi Mountains; inner Coast Range from about Merced County south to San Luis Obispo and Kern counties.

6. *Stipa occidentalis* (fig. 51c) Western Stipa

Densely tufted; culms 25–45 cm. tall; sheath glabrous to pubescent, blade commonly narrow, 1–2 mm. wide; panicles 10–20 cm. long; glumes 10–15 mm. long; lemma 6–8 mm. long, uniformly appressed-hairy; awn twice-bent, densely hairy on all three segments.

Native. Dry slopes, flats and ridges mostly above 5000 feet; higher mountains of southern California north through the Sierra Nevada and Cascades; North Coast Ranges; thence north and east beyond the California border.

7. *Stipa elmeri* (pl. 2c; fig. 53a) Elmer Stipa

Similar to and probably not distinct from *Stipa occidentalis*, differing only in the taller culms and commonly dense, short pubescence of the sheaths and blades. The terminal segment of the awn is merely scabrous, the lower two segments hairy as in *Stipa occidentalis*. Elmer Stipa is more adapted to dry soils on the east slope of the Sierra Nevada and Cascades where it is common.

8. *Stipa californica* (fig. 51b) California Stipa

Culms 60–100 cm. tall; sheaths puberulent about the collar; panicles 20–30 cm. long, narrowly contracted; lemmas 6–8 mm. long, appressed-hairy with longer and more spreading hairs towards the apex; awn twice-bent, scabrous or sometimes hairy on the lower two segments.

Native. Open areas in timber or rocky slopes above 4000 feet Sierra Nevada, Cascades, North Coast Ranges, thence north.

9. *Stipa columbiana* Columbia Stipa

Similar in many respects to *Stipa californica* differing in the glabrous collar and always uniformly scabrous awn. Very often the lower sheaths are reddish-purple colored.

Native. Distribution as *Stipa californica*, sometimes common about the edges of or in meadows.

10. *Stipa lemmoni* (fig. 54) Lemmon Stipa

Densely tufted; culms 30–100 cm. tall; blades flat to involute; panicles 8–15 cm. long, narrow-contracted; glumes broad, the lower 5-nerved, the upper 3-nerved; lemmas 5–7 mm. long, plumpish, pale to brownish, appressed-hairy; callus blunt.

Native. Open woods, edges of brush, from about 2000 to 6000 feet; mountains of southern California north in the Sierra Nevada and Cascades; Coast Ranges north of San Francisco Bay; thence north beyond the California border.

33. Genus *Oryzopsis* Ricegrass

Tufted perennials; blades flat or involute; panicles

commonly open or, in a few species, contracted; spikelets 1-flowered, the floret tough and usually readily deciduous; glumes equal or nearly so; lemmas oval to oblong, becoming hardened at maturity, tipped by a straight, curved, or slightly bent awn which easily falls away.

1. Lemma smooth, 2 mm. long; glumes 3 mm. long........
   1. *O. miliacea*
1a. Lemma hairy, 3 mm. long; glumes 6–7 mm. long........
   2. *O. hymenoides*

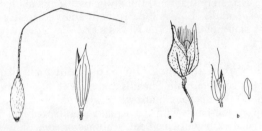

Fig. 54. *Stipa lemmonii*; glumes and floret
Fig. 55. a, *Oryzopsis hymenoides* spikelet; b, *Oryzopsis miliacea*; spikelet and floret

1. *Oryzopsis miliacea* (pl. 5f; fig. 55b) Smilo

Densely tufted perennial, the base rather tough and knotty; culms slender to stoutish, 60–150 cm. tall, commonly widely spreading, solid; blades flat, distributed on the culm; panicles open, 15–30 cm. long, the spikelets borne towards the ends of the long branches; glumes 3 mm. long; florets grayish to brownish, more or less pear-shaped, smooth, about 2 mm. long, the awn straight, slender, about 4 mm. long, early deciduous.

Introduced from the Mediterranean region. Now sometimes naturalized. Smilo has for many years been seeded as a dry land pasture grass and sometimes for erosion control. It has been very useful in brush burn seedings although a good seed bed must be prepared and some patience necessary since the grass seedlings are slow to develop, and they are rather sensitive to strong competition by aggressive annual species. Once established in cleared brush areas the grass persists for many years. It provides good forage for grazing animals and is of use in holding soil in place.

Smilo is seemingly widely adapted in various habitats and occurs primarily at the lower elevations throughout most of California. It sometimes may be found as a weed in vacant lots or in old gardens in some towns or cities. Occasionally the grass occurs along roads and creek channels.

2. *Oryzopsis hymenoides* (pl. 5g; fig. 55a)
Indian Ricegrass

Densely tufted perennial; culms 30–70 cm. tall; blades largely basal, involute, sharp-pointed, sometimes rather firm; ligule prominent to 6 mm. long; panicles 7–20 cm. long, eventually wide-spreading, the branches and pedicels commonly wavy; glumes ovate, abruptly tapering, 6–7 mm long; lemma ovoid to spindle-shaped, commonly dark-colored and long-hairy, the awn hairlike, about 4 mm. long.

Native. In arid or semi-arid regions, sometimes abundant on light or sandy soils; Colorado and Mojave deserts north along the east slope of the Sierra Nevada and Cascades, thence north and east beyond the California border. Indian Ricegrass is a valuable forage for livestock in the deserts, and in many areas the plant has been seriously overgrazed endangering or greatly reducing the stand. Reseeding experiments using the grass have often shown poor results because of unpredictable rainfall and rodent or ant activities. Native Indian tribes used the seeds as food.

34. Genus *Ampelodesmos*
1. *Ampelodesmos mauritanicus* (pl. 5h)

Robust perennial forming fairly large clumps; culms 1–2 meters tall; blades flat, smooth and shiny on the lower surface, serrate along the margin, up to 1 meter in length and up to 7 mm. wide, becoming inrolled upon drying, tapering to a long, fine point; ligule membranous 10–30 mm. long, ciliate along the margin; panicles more or less 1-sided, 30–60 cm. long; spikelets several-flowered, 10–16 mm. long; lemmas 2-toothed at the apex, the midrib extended as a short awn between the 2 teeth, the awn 1–2 mm. long, lemma further long-hairy on the callus and about the midnerve towards the base; palea with the 2 nerves extended as short awns.

Native of northwest Africa. Though attractive as an ornamental the grass is sparingly grown in California. The dry, tan-colored panicle is decorative in dried flower arrangements or quite attractive displayed alone.

35. Genus *Phyllostachys*

Rhizomes woody, long-creeping; culms perennial, woody, D-shaped in cross section, branching at all but the lower nodes, the branches 2 per node, one of the pair branching again from the base and thus appearing as a third branch of the culm node; leaves with blades jointed with the sheaths, the sheaths soon falling away from the culm; blades with rather stiffish bristles at the base; spikelets sessile in the axils of bracts forming a bracteate spike, several-flowered, lemmas about 2 cm. long. Seldom flowering.

Two varieties of a hardy species, *Phyllostachys bambusoides* are commonly grown in California as ornamentals:

a. *Phyllostachys bambusoides* var. *bambusoides*, Giant or Hardy Timber Bamboo, has culms 10–20 meters tall and as much as 5–13 cm. in diameter, the basal nodes spaced.

b. *Phyllostachys bambusoides* var. *aurea* (pl. 6a) Fishpole, Yellow or Golden Bamboo, has culms much shorter, ordinarily less than 5 meters tall, the lowermost nodes densely congested.

36. Genus *Oryza* Rice
1. *Oryza sativa* (pl. 6b) Rice

Aquatic annual; culms 60–100 cm. tall; blades flat to 1.5 cm. wide; ligule membranous, 15–45 mm. long at least on the lowermost leaves; panicle loose, drooping; spikelet with a terminal grain-producing floret with two sterile scales closely appressed about the base; glumes greatly reduced represented by two minute lips below the sterile scales; fertile lemma 7–10 mm long, tough, scabrous, 5-nerved, abruptly pointed at the apex or awned; palea similar to the lemma but narrow and 2-nerved, the nerves nearly marginal. The fertile floret, with the 2 closely appressed sterile scales immediately below, falls as a unit at maturity from the liplike glumes at the tip of the pedicel.

A cultivated crop, germinating and growing during the summer period in standing water. Primarily grown in the Sacramento and San Joaquin valleys.

37. Genus *Ehrharta*
1. *Ehrharta calycina* (pl. 6c) Veldtgrass

Densely tufted perennial; culms numerous, slender, 30–60 cm. tall; blades flat, glaucous, often wrinkled part way along the margin, more or less auriculate at the base, the collar often dark red-purple colored; ligule a prominent membrane with several awnlike teeth at the apex; panicles loose, sometimes very open and at other times narrow-contracted, 8–15 (–25) cm. long; spikelets 5–8 mm. long with two well-developed sterile florets enclosing the upper fertile one, the 3 falling as a unit at maturity; sterile lemma awned or at least short-pointed at the apex; glumes becoming purplish, nearly equal, a bit shorter than the sterile lemmas.

Introduced from South Africa. Now naturalized mostly on light, sandy soils in the central and southern coastal counties of California. Veldtgrass is valuable as a livestock forage and as a soil stabilizer. The grass does not persist in some areas and periodic reseeding is necessary. In grazing areas where good stands are present, some care must be exercized in the management of animals for they "camp" on the tufts and literally graze them out.

38. Genus *Arundo* Giant Reed
1. *Arundo donax* (pl. 6d) Giant Reed

Robust perennial forming a dense network of very thick, scaly rhizomes; culms erect, nearly woody, 2–6 meters tall, 1–4 cm. thick; blades flat, evenly spaced along the culm, 3–6 cm. wide, clasping the stem at the base, strongly scabrous on the margin; ligule a short membrane fringed with very short hairs; panicles often more or less contracted but sometimes varying to open and plumose; spikelets 8–16 mm. long, ordinarily 3-flowered; lemma thin, long-hairy on the lower half, 5-nerved, the nerves more or less prolonged as short awns at the apex.

Introduced from the Old World. Widely planted as windbreak or screen in California, particularly in the rural areas. It is occasionally used as an ornamental in cities and towns, particularly a form with variegated leaves. The culms are useful as stakes, as fishing poles and, to some degree, in the manufacture of reeds for musical instruments.

39. Genus *Phragmites* Reed
1. *Phragmites australis*
(synonym: *Phragmites communis*) (pl. 6)
Common Reed

Perennial by creeping rhizomes; culms erect 1.5–3 meters tall or sometimes stoloniferous 5–20 mm. thick; blades flat, to 3.5 cm. broad; ligule a short and fringed membrane; panicles 8–40 cm. long, 2–8 cm. wide somewhat brownish to purplish; spikelets 12–15 mm. long, several-flowered, the lowest floret sterile or staminate, the rachilla long-hairy; lemma glabrous, the lowermost one the longest then becoming progressively smaller upwards.

A cosmopolitan species; in California the grass occurs in marshes, along sloughs and other waterways, occasional on the margins of lakes or below springs. The culms are useful as stakes for bedding plants. The native Indian tribes probably used the canes in certain baskets or other types of utensils and perhaps as implements or arrow shafts.

40. Genus *Cortaderia* Pampasgrass
1. *Cortaderia selloana* (pl. 6f) Pampasgrass

Robust perennial forming large clumps and mounds of foliage 1–2 meters high; culms to 4 meters tall; blades to 1.5 meters long, flat, the midrib raised on the undersurface, strongly scabrous on the margin, long tapering to a fine point; ligule a fringe of hairs, also hairy on the lining of the sheath just below the ligule; panicles plumelike, silky-white to purplish, as much as 90 cm. long; spikelets 2–3-flowered, unisexual or nearly so, the pistillate spikelets bearing long silky hairs, the staminate spikelets glabrous. The attractive panicles are produced in the late summer or fall.

Introduced from South America where it is native in Brazil, Argentina, and Chile. Widely planted as an ornamental in California, some forms of it producing a little seed and becoming naturalized especially along the coast.

41. Genus *Danthonia* Oatgrass

Tufted perennials; culms erect to widely spreading and all California species produce fertile 1–2-flowered cleistogamous spikelets (cleistogenes) within the basal sheath, the culms easily breaking away from the clump; panicles reduced to a few appressed or widely divergent large spikelets mostly exceeding 1 cm. in length; spikelets of 5–10 florets; glumes longer than the uppermost floret; lemmas rounded on the back, hairy, at least along the margin in the lower half, the apex of the lemma with a deep notch, an acute, usually awned tooth on either side of the notch. A flat, loosely twisted awn arises from the back of the lemma below the notch and is once-bent above the twisted portion, the terminal segment of the awn straight or slightly curved; 3 native species in California.

1. Panicle of 5 or fewer spikelets, if more than 1 then on somewhat divergent to wide-spreading branches.
    2. Spikelets 2–5 in number; foliage glabrous to hairy; culms mostly exceeding 30 cm. in height............
       1. *D. californica*
    2a. Spikelets 1 or 2; sheaths densely long-hairy; culms less than 30 cm. high..................2. *D. unispicata*
1a. Panicle narrow, of 8–12 closely appressed spikelets......
    3. *D. pilosa*

Fig. 56. *Danthonia californica*; a, habit; b, spikelet; c, floret

1. *Danthonia californica* (fig. 56) California Oatgrass

Densely tufted; culms 40–80 cm. tall; sheaths and blades ordinarily hairless excepting tufts of long hairs on either side of the collar; panicles commonly reduced to 3 widely divergent spikelets; glumes variable, 1–2.5 cm. long, exceeding the florets; lemmas hairy along the margin in the lower half and about the callus.

The var. *americana*, American Oatgrass, is a less robust variant of the species differing primarily in the hairy sheaths and blades. It replaces the typical at medium to high elevations.

Native. Coast Ranges and Sierra Nevada, thence north to Canada and east to the Rocky Mountains.

The typical California Oatgrass and its variety are valuable livestock forage. Grazing animals seek out the individual plants and consistently overgraze them, sometimes causing rapid depletion of good stands. It prefers moist soils or at least soils with adequate subsurface moisture. Arid soils do not support these grasses at all, consequently they are generally absent from the hills of the Inner Coast Range and Sierra Nevada foothills. The typical is best developed along the north coast of California and the variety in the mountains above 2000 feet.

2. *Danthonia unispicata* Onespike Oatgrass

Low, densely tufted perennial similar to, and possibly not specifically distinct from *Danthonia californica* var. *americana*. The most significant features of Onespike Oatgrass are: the consistently low stature of the plant, the culms rarely exceeding 25 cm. in height, the exceptionally dense long-hairy sheaths and 1 or

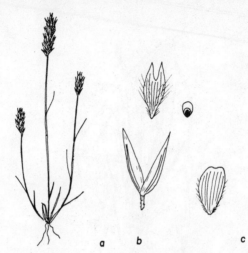

Fig. 57. *Schismus*; a, *Schismus arabicus* habit; b, glumes, floret and grain of *Schismus arabicus*; c, *Schismus barbatus* floret

sometimes 2 spikelets at the tip of the stem.

Native. Rocky slopes, flats and dry meadows, commonly associated with open coniferous forests particularly in the Sierra Nevada and Cascades, though extending into the North Coast Ranges. It is a valuable forage and erosion control plant.

3. *Danthonia pilosa* Hairy Oatgrass

Loosely tufted perennial; culms slender to 60 cm. tall; foliage rather loosely hairy about the base; panicles narrow, consisting of 8–12 closely appressed spikelets; glumes 10–14 mm. long enclosing 6–7 florets; lemmas hairy on the back near the base with tufts of hairs along the margin, callus well developed, hairy.

Introduced from Australia. Naturalized mostly along the north coast in Humboldt, Mendocino and Sonoma counties, with scattered localities south to Ventura County and perhaps north into Oregon.

Hairy Oatgrass is wiry and relatively unpalatable to sheep or cattle. It grows in colonies and develops to such an extent in some areas as to crowd out better forage species, so it is regarded as a range pest.

42. Genus *Schismus* Schismus

Low annuals with slender culms and short blades; ligule a fringe of hairs; collar with tufts of long hairs on either side; panicles contracted to almost spikelike; spikelets 5–10-flowered, the rachilla usually disjointing

Fig. 58. *Aristida oligantha* habit

between the florets; glumes 3–7-nerved and about as long as all the florets; lemmas 7–9-nerved, rounded on the back, commonly hairy along the margin in the lower half, the apex 2-lobed (bifid); grain shiny, transparent, loosely enclosed between the lemma and palea.

There are 2 naturalized species rather widespread in the arid climates of California. Both are common from the central San Joaquin Valley south to San Diego County and east through the Mojave and Colorado deserts.

Schismus are short-lived, rapidly growing from seed, flowering and soon shattering away, life span dependent upon sufficient but usually small amounts of rainfall. They are of value temporarily in checking soil erosion and as sheep forage.

1. Glumes 5–6 mm. long; lemmas 2.5–3 mm. long, the lowermost with a deep notch between 2 acute but short lobes, lemma margin rather long-hairy below..*Schismus arabicus*
Arabian Schismus (figs. 57a, 57b)
1a. Glumes 4–5 mm. long; lemmas about 2 mm. long, the apex of the lowermost only shallowly notched with scarcely noticeable rounded lobes on either side; lemma margin rather sparsely short-hairy............*Schismus barbatus*
Mediterranean Schismus (fig. 57c)

43. Genus *Aristida* Threeawn

Tufted perennials or annuals; blades commonly slender, tightly rolled; ligule a fringe of hairs; panicle contracted to somewhat loose and open; spikelets 1-flowered; glumes persistent, thin, long-tapering into a fragile, often deciduous awn; lemma tough and leathery, scabrous, cylindric, sometimes purple-blotched, tapering at the apex into a straight and sometimes twisted column supporting a 3-partite awn (in California species), the awn segments ascending or widely divergent; palea thin and hyaline; callus sharp-pointed, densely short-hairy.

A large genus of some 330 species spread over the temperate and subtropical regions of the world typically in arid or semi-arid regions. California has 12 species. The forage value of the Threeawns is often poor but in the absence of other palatable forage as in the deserts, they assume some importance as forage. The two species described below are weedy but are the most common.

1. Awn segments 4–7 cm. long, widely divergent............
   1. *A. oligantha*
1a. Awn segments mostly less than 2 cm. long..............
   2. *A. adscensionis*

Fig. 59. *Aristida oligantha* floret

1. *Aristida oligantha* (figs. 58, 59) Prairie Threeawn

Annual developing during the summertime; culms wiry, commonly branching well above the base, 20–50 cm. tall; spikelets

Fig. 60. *Aristida adscensionis*; a, habit, b, glumes, floret

large, more or less scattered along the culm on short pedicels or on short branches; glumes 2–3 cm. long, short-awned at the tip; floret 1, firm, cylindrical, grayish, often red-purple blotched, sharp-pointed at base, the upper portion tapering into 3 widely divergent awns.

Introduced from the eastern half of the United States. Now widely naturalized in California from San Diego County north to Siskiyou County; occurs in valleys and foothills but does not ascend into the mountains much above 3500 feet; frequent in ditches along roads or low spots in fields. It is curious that this grass is able to grow on dry soils, thrive, and flower during our hot summers when most other annuals have long since dried up. Prairie Threeawn is valueless as forage and when abundant is considered a range weed.

2. *Aristida adscensionis* (fig. 60) Sixweeks Threeawn

Annual; culms branching at the base and also commonly above hence becoming "bushy"; culms 10–50 cm. tall; panicles narrow and compact by close clusters of spikelets 5–10 cm. long; glumes unequal, the upper 8–10 mm. long, the lower shorter; floret 6–9 mm. long, the three awns at the tip about equal, 10–15 mm. long.

San Luis Obispo County and Tehachapi Mountains south to San Diego and Imperial counties; thence to Texas and south to Argentina.

44. Genus *Eragrostis* Lovegrass

Annual or tufted perennials; the blades with a tuft of long hairs on either side where it joins the sheath; ligule a fringe of short hairs; panicles open, sometimes large and diffuse; spikelets often lead-gray, dark olive-green or sometimes pale green, rarely purplish-colored; florets several to many in each spikelet, the lemmas 3-nerved and ordinarily falling away from the rachilla with the ripened grain loosely enclosed, the palea curved, more or less adherent to the rachilla.

1. Glandular plants.........................1. *E. cilianensis*
1a. Nonglandular plants.
    2. Panicles dark-colored; grains dark brown, shallowly furrowed along one side............2. *E. orcuttiana*
    2a. Panicle pale green to purplish, rarely dark-colored; grains light brown, not at all furrowed....3. *E. diffusa*

1. *Eragrostis cilianensis* (fig. 61) Stinkgrass

Summer-flowering annual; culms decumbent or erect 10–35 (rarely to 50) cm. tall, with small glandular areas below the nodes, on the foliage and, on the inflorescence; panicle 7–20 cm. long, the branches rather short, bearing numerous spikelets on short pedicels; spikelets 7–15 mm. long, 2.5 to 3 mm. wide 10 to as much as 40-flowered; lemmas about 2.5 mm. long with minute glands along the midrib; grains dark brown, roundish, minute.

Introduced. Weedy in areas of summer irrigation at lower elevations throughout California. The numerous glands impart a peculiar odor to this plant accounting for the common name. Large plants produce many thousands of fine seeds.

Fig. 61. *Eragrostis cilianensis*

2. *Eragrostis orcuttiana* (fig. 62a)
Orcutt Lovegrass

Summer-flowering annual; culms 40–100 cm. tall; panicles open, 15–30 cm. long; spikelets usually dark-colored, grayish-black, or deep olive-green, 6–10-flowered, 5–7 mm. long, about 1 mm. wide; grain dark brown, shallowly furrowed along one side, oblong, slightly more than 1 mm. long.
Fig. 62 about here

Native. A summer weed in irrigated areas at low to medium elevations throughout much of the state. It is sometimes quite abundant in various croplands. The large production of seed ensures a good stand of the grass the succeeding year.

3. *Eragrostis diffusa* (figs. 62b, 63)
Spreading Lovegrass

Summer-flowering annual; culms ordinarily numerous, 20–50 cm. tall; panicles diffuse bearing numerous spikelets rather closely appressed to the branches; spikelets 5–8 mm. long, 1.5–2.0 mm. wide; grains light brown, oblong, slightly more than 1 mm. long, not at all furrowed along one side.

Native. Common as a weed in summer irrigated areas at low elevations in California, and in orchards, vineyards, various croplands, pastures, and ditches along roads. Large plants produce several thousand seeds.

45. Genus *Sporobolus* Dropseed

Perennials or annuals; blades flat or involute, most of them about the base; panicle open to contracted; spikelets grayish, the single floret readily falling from between the glumes; glumes unequal, ordinarily one of them as long as the lemma; lemma awnless, thin, 1-nerved; grain loosely enclosed between the lemma and palea and often falling free at maturity, accounting for the common name.

1. Panicle ordinarily diffuse, often partially included at the base.....................................1. *S. airoides*
1a. Panicle ordinarily included in the upper leaf sheath, sometimes partially exserted...............2. *S. cryptandrus*

1. *Sporobolus airoides* (Cover-B; fig. 64)
Alkali Sacaton

Perennial forming tough clumps; culms 50–100 cm. tall, shiny and cartilaginous at the base; blades flat to involute; sheaths ordinarily short-hairy, sometimes long-hairy at the summit; ligule a fringe of hairs; panicle diffuse, often partially included in the expanded upper sheath; lemma about 2 mm. long, smooth, obtuse.

Native. On alkaline flats from Solano County south through the Central Valley to southern California, thence east through the deserts. The grass provides good grazing for cattle and horses when locally abundant. It was a common element of the original California grassland, but intensified agriculture in the San Joaquin Valley has removed much of the grass.

2. *Sporobolus cryptandrus* (fig. 65) Sand Dropseed

Perennial forming small to medium tufts; culms 30–70 cm. tall;

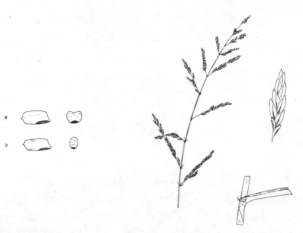

Fig. 62. Grains of *Eragrostis* greatly enlarged and viewed from the side and in cross-section; a, *Eragrostis orcuttiana*; b, *Eragrostis diffusa*

Fig. 63. *Eragrostis diffusa* showing the panicle, a spikelet, and a portion of a leaf

sheaths always with a conspicuous tuft of white hairs at the summit on either side; panicles commonly completely included in the elongate upper sheath, sometimes the upper portion of the panicle exserted and expanded; lemmas 2–2.5 mm. long.

Native. On light, sandy soil at various and often distant localities throughout much of California from low to high elevations; also east and south beyond the state borders; common along roadsides in some areas.

46. Genus *Crypsis* Pricklegrass
1. *Crypsis schoenoides*
(synonym: *Heleochloa schoenoides*)
(fig. 66) Swampgrass

Annual; flowering during the summer and fall; culms 5–50 cm. long, largely decumbent; sheaths dilated, loose on the culm; blades flat with slender tips, usually less than 10 cm. long; panicles very dense, pale to purplish, 2–5 cm. long, nearly 1 cm. thick, partially included in the swollen uppermost sheath; spikelets about 3 mm. long, the glumes shorter than the single floret; lemma and palea loosely enclose the grain.

Introduced from Europe. Now naturalized throughout most of California; common in bottomlands that are inundated with water during the winter and spring, along stream channels, and

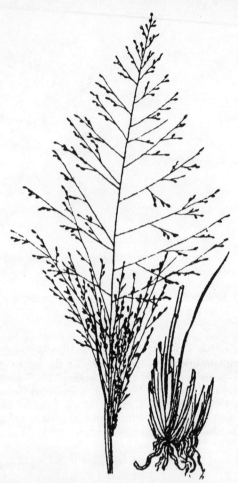

Fig. 64. *Sporobolus airoides*

about reservoirs either along the edge or on the beds when the water evaporates or is used. After the water recedes Swampgrass germinates on the mud, forming at first a short cluster of leaves. Further drying of the mud and increasing temperatures stimulate elongation of the stems and eventually the panicles emerge by late summer or fall. It is an esteemed forage for wildfowl and is usually abundant about the wildfowl refuges. Livestock graze the grass around the edges, or on the dry beds of, reservoirs or when it is abundant in stream channels.

Old specimens of the grass that have developed on rich soils may produce as many as 100 prostrate stems forming a mat at

Fig. 65. *Sporobolus cryptandrus*; spikelet, floret and habit
Fig. 66. *Crypsis schoenoides* showing the habit of the plant, a spikelet and the grain

least 3 feet in diameter. On poor sites only 3 or 4 stems may develop and the plants are scarcely over 4–6 inches in diameter. In either case the large number of spikelets in the panicles ensure large production of seed and a good stand of the grass for the next growing season.

47. Genus *Muhlenbergia* Muhly

Perennials or a few annuals, the perennials commonly with tough, scaly rhizomes, these often close and intertwined giving a knotty appearance to the base of the plant; blades flat or tightly rolled; ligule a membrane that, in some species, is also fringed with hairs at the apex; panicle open, loosely contracted or spikelike; spikelets 1-flowered; glumes unequal, mostly shorter than the lemma; lemma 3-nerved, often lead-colored, smooth or hairy, awned or awnless.

Sixteen species grow in California, but none occurs in any abundance over wide areas. In some local habitats certain Muhly species may be abundant. As forage grasses they are generally good but are of only secondary importance. The strong root system and the rhizomatous character of the perennials make them valuable in local erosion control.

1. Lemma long-awned, the awn 1 cm. or more long........
   1. *M. microsperma*
1a. Lemma awnless or very short-awned.
    2. Robust coarse plant; culms 60 cm. or more tall......
       2. *M. rigens*
    2a. Low plants; culms less than 60 cm. tall.
        3. Culms slender; plants averaging about 10 cm. tall.
           3. *M. filiformis*
        3a. Culms wiry; plants averaging about 25 cm. tall...
            4. *M. richardsonis*

1. *Muhlenbergia microsperma* (fig. 67)
Littleseed Muhly

Annual; culms slender, 10–45 cm. tall, commonly purplish-red, branching at or above the base; foliage usually reddish, blades flat, those on the culm 2–3 cm. long; panicles narrow, the branches spreading but short, 5–15 mm. long, or occasionally the lower to 20 mm. long in robust specimens; spikelets on short,

Fig. 67. *Muhlenbergia microsperma* spikelet and a single branch

stubby pedicels; glumes minute; lemma 2.5–3.5 mm. long, tipped by a slender awn 10–20 mm. long. The florets are readily deciduous from the panicle and in dried specimens nearly all of

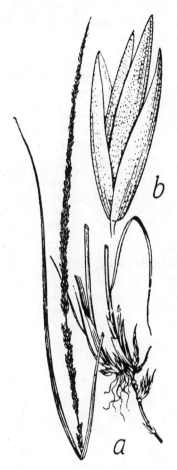

Fig. 68. *Muhlenbergia rigens;* a, habit; b, spikelet

them fall away from the glumes. The species produces usually a single spikelet in the axils of the lower leaves, the enveloping sheath tightly enclosing it. It matures a somewhat plumper grain than those spikelets of the panicle.

Native. On moist to dry soils along streams, among shrubs, along roads, slopes, rocky ridges, and sometimes rather common on disturbed soils, from Monterey and Kern counties south to Lower California. It is occasional in areas beyond this range where it was probably introduced.

2. *Muhlenbergia rigens* (pl. 6g; fig. 68) Deergrass

Coarse, perennial, densely tufted from a knotty, close-rhizomatous base; culms 60–150 cm. tall, erect to widely spreading in the clump; blades scabrous and ordinarily elongate 20–50 cm. long and long-tapering to a fine tip; panicles whiplike, dense, 15–50 cm. long; spikelets numerous, grayish; lemmas 3–3.5 mm. long, awnless, though sometimes with a short point at the tip.

Native. Along streams, edges of meadows, seeps on hillslopes, ditches and roads from low elevations to about 7000 feet; San Diego County north in the Coast Ranges to the Monterey Bay Area, and in the Sierra Nevada to Shasta County; occasional in the Central Valley.

Deergrass is summer or fall flowering, the older and coarse clumps scarcely if at all palatable to livestock, although sometimes cattle and horses graze the tender new foliage of the younger clumps but soon pass on to more succulent plants. The grass is readily identified by its size and the long whiplike panicle.

3. *Muhlenbergia filiformis* (fig. 69)
Pullup Muhly; Slender Muhly

Annual; or apparently when low and mat-forming, a short-lived perennial; culms slender, 5–15 cm. tall; blades 1–3 cm. long, flat but narrow; panicles contracted, slender, 2–5 cm. long; glumes

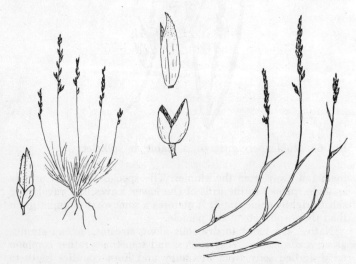

Fig. 69. *Muhlenbergia filiformis* habit and spikelet
Fig. 70. *Muhlenbergia richardsonis* spikelet and three branches

ovate, about 1 mm. long; lemma grayish-green, about 2 mm. long, usually abruptly pointed or sometimes shortly awned.

Native. Mostly in meadows or along streams or occasionally open woods in moist spots, all locations above about 3500 feet; throughout most of California and extending north and east beyond the borders. It is fairly useful forage for sheep, though good stands are limited.

4. *Muhlenbergia richardsonis* (fig. 70) Mat Muhly

Perennial, ordinarily forming dense clumps from a close-rhizomatous base; culms wiry 10–40 cm. tall; blades flat or rolled; panicles slender, almost spikelike, 2–10 cm. long; glumes ovate, about half the length of the lemma; lemmas 2–3 mm. long, acute to abruptly pointed.

Native. Dry to moist mountain slopes, flats or valleys, edges of meadows, mostly in open areas above 5000 feet in elevation; higher mountains of southern California, north through the Sierra Nevada and Cascades to Washington, also east to the Rocky Mountains and beyond.

The grass is of local forage value along the east slopes of the Sierra Nevada where it is best developed between 6000–9000 feet. It recovers rather well from close grazing early in the growing season.

48. Genus *Chloris*
Fingerggrass; Chloris; Windmillgrass
1. *Chloris virgata* (fig. 71)
Showy Chloris; Feather Fingerggrass

Annual, flowering during summer or fall; culms 30–60 cm. tall, branching at the base or above, when the latter then rather "bushy"; sheaths dilated, loose on the culm, blades flat; spikes several 2–8 cm. long, erect, usually whittish, feathery or silky; spikelets densely crowded along one side of the spike, with 2 florets, the lower one grain-producing, long-hairy on the margin near the apex, provided with a hairlike awn 5–10 mm. long at the apex; upper floret sterile, somewhat triangular in outline, with a hairlike awn. The 2 florets fall away as a unit from the glumes.

Showy Chloris is a rather common weed of roadsides, ditches, pastures, or any area where there is summer irrigation; mostly at low elevations throughout California. The attractive inflorescence suggests it may be useful in dry bouquets, but the florets readily fall away unless collected early.

Fig. 71. *Chloris virgata*; a, inflorescence; b, glumes; c, florets

49. *Cynodon* Dogtoothgrass
1. *Cynodon dactylon* (fig. 72) Bermudagrass

Perennial by extensive creeping rhizomes and stolons; the erect or flowering stems 10–30 cm. tall; ligule a fringe of hairs; sheath with a tuft of hairs at the summit on either side of the stem however, in some forms, the sheaths and blades sparsely hairy; spikes digitately arranged at the summit of the stem, 4 to 7 in number, slender, 2–5 cm. long; glumes shorter than the floret, persistent on the spike; lemmas boat-shaped, about 2 mm. long, short-hairy on the keel and the marginal nerves.

Introduced from the Old World. Now naturalized as a formidable weed of ornamental plantings and croplands, the extensive network or rhizomes and stolons being difficult to eradicate fully. Despite its weedy character in some areas, it is still useful as pasture, as lawn, and in holding soil in place. There is some objection to the grass as a lawn because it browns during the winter. Infestations in Bluegrass and Bentgrass lawns show up as unsightly patches in cold weather. In yet other areas Bermudagrass is valued as a durable lawn, despite its winter appearance, since it will tolerate soils where other grasses would do poorly.

It is common along roads, highways and may be found along

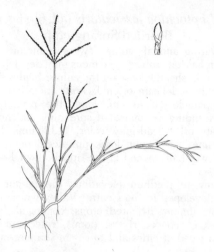

Fig. 72. *Cynodon dactylon*; habit and a spikelet

sidewalks or in vacant lots in cities and towns. Bermudagrass is absent on summer-dry soils of the valley plains and foothill slopes.

50. Genus *Leptochloa* Sprangletop

Annuals (California species) or some tufted perennials; culms erect or geniculate below, normally branching from the base and from the nodes above as well, the whole plant dense and bushy; blades, in those species treated here, strongly scabrous; inflorescence a panicle of usually numerous spikes or racemes branches more or less scattered along the upper portion of the culm; spikelets several-flowered, essentially borne along one side of each spike; florets readily breaking apart and falling to the ground; the thin glumes persistent on the rachis; lemma 3–nerved, short-awned or awnless, the lateral pair placed near the margins and these often pubescent at least toward the base.

1. Lemmas awned........................1. *L. fascicularis*
1a. Lemmas awnless......................2. *L. uninervia*

1. *Leptochloa fascicularis* (pl. 6h; fig. 6)
Bearded Sprangletop

Summer-flowering annual; culms few or numerous, 20–75 cm. tall, in certain habitats robust specimens may develop with culms over 1 meter tall; sheaths loose on the culms; blades flat or loosely rolled; ligule a delicate often lacerate membrane 2 mm. or more long; panicle 10–30 cm. long, the primary branches (racemes) ascending or somewhat spreading and up to 10 cm. long; spikelets of a lead-gray color, 6–12 mm. long, 6–12-flowered, lemmas 3.5–5 mm. long, pubescent on the nerves at least in the lower half, awned at the tip, the awn 1-several mm. long.

Native. Low to medium elevations throughout California, though best developed in the Central Valley where it is a common weed of summer irrigated crops; occurs along roads, in ditches, edges of creeks, rivers, ponds, reservoirs and lakes; occasionally a weed of cities and towns in summer-moist places.

2. *Leptochloa uninervia* (pl. 7) Mexican Sprangletop

Summer- and fall-flowering annual of the same habit at *Leptochloa fascicularis* differing, primarily, in the much darker-colored, almost blackish spikelets and awnless lemmas.

Native. Through most of southern California and extending north in the Coast Ranges and Central Valley; weedy along roads, irrigation ditches, about corrals and pastures, and generally in summer irrigated crops.

51. Genus *Hilaria* Hilaria
1. *Hilaria rigida* (fig. 73) Big or Woolly Galletagrass

Coarse rhizomatous perennial; culms solid, densely close-woolly, 45–80 cm. tall, commonly branching above the base; blades stiffish, sharp-pointed; inflorescence a terminal spike 4–8 cm. long, the rachis wavy; spikelets in clusters of 3, the lateral pair of spikelets staminate the glumes of which are several-awned, the central spikelet grain-producing the glumes of which are divided into several narrow segments and awns; the cluster of 3 spikelets readily falls away from the rachis.

Native. In the Mojave and Colorado deserts thence east to Utah and south into Mexico; mostly in sandy and gravelly washes and on alluvial fans. The "brushy" character and woolly stems makes Big Galletagrass readily identifiable even without spikelets. It is a valuable forage grass, usually always grazed down to the ground. Successive years of close grazing have removed many stands that are now represented only by scattered colonies of the grass.

Fig. 73. *Hilaria rigida*; a, habit; b, spike

52. Genus *Distichlis* Saltgrass
1. *Distichlis spicata* (pl. 7b, c) Saltgrass

Perennial by extensive creeping, yellowish, scaly rhizomes forming large colonies; blades rather stiffly 2-ranked along the culms, flat or loosely rolled, usually bluish-green; ligule a short fringe of hairs; panicles reduced to a few short branches and relatively few large, several- to many-flowered spikelets; spikelets purplish to straw-colored, those in each panicle all unisexual however both sexes have similar spikelets; staminate panicles are usually borne well above the leaves, the pistillate panicles equaling or shorter

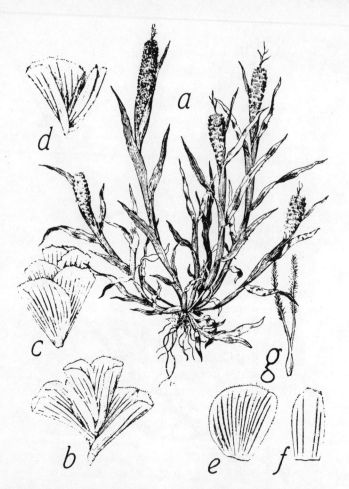

Fig. 74. *Neostapfia colusana*; a, habit; b, c, spikelets; d, floret; e, lemma; f, palea; g, pistil

than the blades; lemmas firm smooth, 5–9-nerved, 3–6 mm. long.

Native. Alkaline or saline flats from the seacoast to the interior, throughout North America to South America. Saltgrass is valuable as a range forage often occurring on sites where there is a scarcity of other vegetation. It provides green forage over most of the year and recovers well from close grazing and trampling by animals.

There are various growth forms within the species: some are coarse-leaved and coarse-stemmed, others stoloniferous, some are fine-leaved (var. *nana*) and on scarcely alkaline soils. Colonies of Saltgrass may be either wholly staminate or wholly pistillate; where there is a mixture of sexes good seed crops are produced.

53. Genus *Neostapfia* Neostapfia
1. *Neostapfia colusana* (pl. 7d; fig. 74) Neostapfia

Glandular gray-green annual; culms ordinarily decumbent, 5–40 cm. long; foliage with numerous raised glands along the margin and nerves; sheath loosely folded about the culm there being no line of demarcation between it and the blade; ligule wanting; panicles pale, dense, cylindric, 2–7 cm. long, the axis prolonged beyond the spikelets and with or without small bracts; spikelets 6–7 mm. long, ordinarily 5-flowered, the lemmas many-nerved, fan-shaped, about 5 mm. long and fringed with very short hairs; glumes wanting; grain obovate, dark brown, viscid.

Native. Occurs only in the Central Valley of California on the dry beds of large vernal pools or vernal lakes. It is one of the rarest grasses in California because the proper habitat for it is uncommon. It is known to occur in Solano and Stanislaus counties, but in the Merced and Colusa (the type locality and source of the specific name) county sites the grass may no longer be present. It is included here because it is the most unusual grass in the state.

54. Genus *Panicum* Panicum

Annuals, or mostly perennials, of various growth habit; culms branching from the base and commonly above as well; ligule a fringe of hairs; blades flat, broad, glabrous to hairy; inflorescence typically an open panicle, though in some narrow and contracted; spikelets plumpish, consisting of a terminal grain-producing floret and directly below a sterile floret consisting of a glumelike lemma and sometimes with a thin palea as well and then, sometimes, with stamens enclosed; glumes very unequal, the upper one similar to and as long as the sterile lemma; fertile floret tough, commonly ovatish, shiny, the lemma margins inrolled over the enclosed palea.

A large genus with about 150 species in the United States but relatively few (10 species) in California. Aside from several introduced weedy species, the native ones are widely distributed but seldom occur in dense stands.

1. Perennial, with a basal rosette of leaves.................
   1. *P. lanuginosum* var. *fasciculatum*
1a. Annual, never forming a basal rosette of leaves.
   2. Plants coarsely hairy; panicles diffuse and readily breaking away from the culm below at maturity.
      3. Fertile floret without a conspicuous basal scar..
         2. *P. capillare*
      3a. Fertile floret with a crescent-shaped basal scar..
         3. *P. hillmanii*
   2a. Plants glabrous or nearly so; panicle not breaking away from the culm................4. *P. dichotomiflorum*

1. *Panicum lanuginosum* var. *fasciculatum*
(*Panicum pacificum*) (fig. 75) Pacific Panicum

Tufted perennial developing, during the spring, a basal rosette of short, broad, densely hairy leaves and simple culms with a well-elevated terminal panicle. During the summer and fall the culms that developed during the spring freely branch above the base with greatly reduced panicles borne among the leaves, the whole becoming quite "bushy." Plants 30–60 cm. tall, the primary panicle 5–10 cm. long, the rachis hairy; spikelets about 2 mm. long, the glumes and sterile lemma hairy.

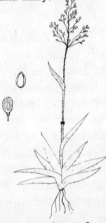

Fig. 75. *Panicum lanuginosum* var. *fasciculatum* habit, spikelet and floret

Native. Moist soils below 5000 feet; San Diego County north in the southern California mountains, Sierra Nevada and Coast Ranges, thence to Canada and east across the United States. The grass occurs along streams, in meadows, bogs, seeps on hillslopes, roadbanks, edges of reservoirs and ponds along irrigation ditches and moist areas among brush.

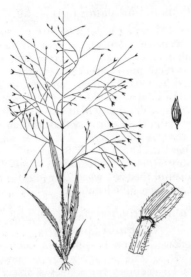

Fig. 76. *Panicum capillare* showing habit of the plant, a spikelet and a portion of a leaf

2. *Panicum capillare* (pl. 7e; figs. 76, 77a)
Common Witchgrass

Summer-flowering annual; culms usually numerous and widely spreading, 20–60 cm. tall; foliage commonly coarsely hairy; panicle diffuse, sometimes as much as half the length of the culm, the whole panicle eventually breaking away and behaving as a tumbleweed in the wind; spikelets glabrous 2–3.5 mm. long, the upper glume and sterile lemma acuminate at the apex.

Native or perhaps some forms introduced from the eastern half of the United States. In California the grass is weedy on the summer-irrigated soils of croplands, pastures, drainage ditches along the road or margins of reservoirs and ponds. Sometimes it forms dense stands locally.

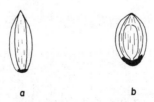

a   b

Fig. 77. Grain-bearing florets showing the nature of the basal scar; a, *Panicum capillare*; b, *Panicum hillmanii*

3. *Panicum hillmanii* (pl. 7f; fig. 77b)
Hillman Panicum

Resembles *Panicum capillare* but is coarsely hairy, has stiffer blades and panicle branches and the fertile floret is dark-colored with a prominent crescent-shaped scar at the base. The grass grows on much drier soils than does *Panicum capillare*.

Introduced from Kansas, Oklahoma or Texas where it is native. Now naturalized in California where it is assuming some prominence along roads in the Central Valley and surrounding foothills. The airy panicles break away and behave as tumbleweeds in the wind often traveling great distances and scattering spikelets along the way. Hillman Panicum is resistant to certain herbicides used to control roadside weeds hence is conspicuous when other plants have been killed out.

4. *Panicum dichotomiflorum* (fig. 78)
Fall Panicum

Summer or fall flowering annual; culms erect or more often widespreading, 25–80 cm. tall; foliage without hairs except at the collar region and sometimes the blades sparsely hairy on the upper surface; lower sheaths loose about the culm often reddish purple, sometimes pale green; panicle open, 10–30 cm. long; spikelets 2–3 mm. long.

Fig. 78. *Panicum dichotomiflorum* panicle and two views of a spikelet

Introduced from the eastern United States. Now naturalized in California mostly as a weed in or about summer irrigated croplands and pastures; sometimes at the edges of reservoirs or in streambeds, drainage ditches or on flats along roads. Fall Panicum is fairly common in the Central Valley but is also intermittent at the lower elevations of the Sierra Nevada, Coast Ranges, and in southern California.

55. Genus *Digitaria* Crabgrass

Annuals (California species); culms erect to decumbent or even prostrate, branching at, and commonly above the base; spikelets sessile to short-pedicelled, in 2's and 3's alternating in two rows along one side of an angular or sometimes ribbonlike rachis thereby forming a slender raceme; racemes several, radiating from the tip of the stem (digitate) with oftentimes scattered racemes below the terminal whorl; spikelets lanceolate to elliptic, one of the glumes greatly reduced.

1. Plants hairy..........................1. *D. sanguinalis*
1a. Plants glabrous or nearly so............2. *D. ischaemum*

1. *Digitaria sanguinalis* (pl. 7g; fig. 79a)
Large or Hairy Crabgrass

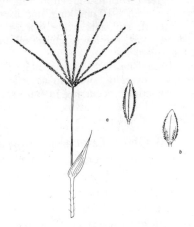

Fig. 79. a, *Digitaria sanguinalis* inflorescence and spikelet; b, *Digitaria ischaemum* spikelet.

Summer-flowering annual; culms 15–60 cm. tall (or long); foliage commonly gray-green but may be green in color, rather coarsely and densely hairy; racemes 5–15 cm. long in a terminal whorl or sometimes an additional whorl below; spikelets lanceolate, about 3 mm. long, the lower glume minute, the upper one hairy along the margin and about half as long as the spikelet.

Introduced from Europe. Now naturalized throughout most of the state, primarily at lower elevations where there is sufficient summer moisture. It is a serious weed of lawns, because the stems are more or less prostrate and usually below the mower blade. The rapid growth of the seedlings soon crowds out the preferred finer-leaved grass species that make up the turf and mats of broad-leaved, gray-green grass spot the lawn and later, following the first frosts, these mats turn brown. Crabgrass is also a weed of orchards, vineyards and other crops, among shrubs or ground cover in ornamental planting, and along roads and ditches. Dense or extensive stands could be cut for hay or pastured.

The seed germinates in the spring or early summer and produces very vigorous seedlings with broad flat blades close to the ground, soon followed by extensive lateral stem development (pl. 7g). Flowering occurs by July or August and continues up to the first frosts thereby producing an abundance of seed.

2. *Digitaria ischaemum* (fig. 79b) Smooth Crabgrass

Summer-flowering annual, similar in growth habit to *Digitaria sanguinalis* except the foliage is essentially hairless and varying from yellowish-green to bluish-green and in age sometimes reddish; 2–6 racemes from 4–10 cm. long, not disposed in a terminal whorl but scattered; spikelets ovate, short-curly hairy on the margin and lower part, the lower glume minute, the upper one as long as the spikelet.

Introduced from Eurasia. Now naturalized through most of the state at low elevations, often in association with *Digitaria sanguinalis* or sometimes occurring in pure stands. It is a weed of summer-irrigated soils: in lawns, ornamental plantings, croplands, roadsides, and ditches.

56. Genus *Eriochloa* Cupgrass

Annual (California species); culms erect or decumbent, often branching above the base; blades flat, smooth, scabrous or hairy; ligule a fringe of hairs; inflorescence of several to many appressed or spreading racemes usually closely spaced on the culm axis; spikelets on short pedicels or sessile along one side of the raceme, more or less hairy, readily shattering away. The most distinctive feature of the genus is the usually dark-colored, thickened callus at the base of the spikelet. The lower glume surrounds the base of the callus like a cup hence the common name.

1. Racemes ascending to somewhat divergent; foliage glabrous to somewhat hairy....................1. *E. gracilis*

1a. Racemes closely appressed to the main axis; foliage ordinarily densely pubescent..................2. *E. contracta*

1. *Eriochloa gracilis* (fig. 80) Southwestern Cupgrass

Summer flowering annual; culms 30–90 cm. tall; leaves commonly without hairs though sometimes the lower sheaths and blades are hairy; racemes 2–3 cm. long, ascending to somewhat spreading; spikelets green or purplish, 4–5 mm. long, abruptly pointed at the apex, softly pubescent, the fertile floret with a short awn less than 1 mm. long.

Native. Central Valley to southern California mostly as a weed in summer irrigated lands as orchards, vineyards, pastures, cotton, and other crops. It does provide some forage for domestic animals when in pastures but mostly before it flowers. The grass is killed out with the first fall or winter frosts.

Fig. 80. *Eriochloa gracilis*; spikelet, fertile floret and the inflorescence

2. *Eriochloa contracta* Prairie Cupgrass

Similar to *Eriochloa gracilis* in growth habit but differing in the ordinarily dense pubescence of the foliage, the closely appressed racemes, and the longer awn on the fertile floret which reaches 1 mm. in length.

Native in southern California; introduced into the Central Valley. It occurs mostly as a weed in summer irrigated crops but occasionally is found along roadside ditches.

57. Genus *Echinochloa* Cockspur

Summer-flowering annuals (California species); blades flat; ligule absent, at least in those described below; spikelets variously hairy, sometimes stiffly so, short-

Fig. 81. *Echinochloa crusgalli*

pedicelled and crowded along one side of the raceme rachis; racemes several scattered along the upper part of the stem; spikelets with the first glume about half the length of the spikelet, the second (upper) glume equaling the length of the spikelet and the sterile lemma; fertile floret planoconvex, smooth, shining.

1. Racemes more than 2 cm. long; ordinarily tall, erect plants. . . . . 1. *E. crusgalli*
1a. Racemes 1–2 cm. long; commonly low plants, the culms often decumbent . . . . 2. *E. colonum*

1. *Echinochloa crusgalli* (fig. 81)
Barnyardgrass; Watergrass

Commonly stout annual; culms erect 30–120 cm. tall; leaves mostly without hairs, except sometimes the sheaths hairy about the base; inflorescence erect or nodding, 10–20 cm. long, greenish to purplish; spikelets about 3 mm. long, numerous in 3 or 4 rows along each of the several racemes; sterile lemma usually awned, the awn quite variable in length, though, in some forms, awnless.

Widely naturalized in the state or perhaps some forms native. Barnyardgrass is perhaps the commonest of the weedy grasses of the summer period. It occurs in all habitats having summer irrigation. In and around the rice fields of the Central Valley it is a weed of major importance. The grass is also common along roads and roadside ditches. It is an important weed of lawns and other ornamental plantings in cities and towns.

As forage the grass is good but the best stands of it are seldom available for animal grazing. It is too succulent to make good hay. The seeds are highly esteemed by ducks and geese as well as other birds. The first frosts of fall or winter effectively kill out the Watergrass plants but not before an abundance of seed has been shed upon the ground.

2. *Echinochloa colonum* (fig. 82) Jungle Rice

Annual; culms commonly decumbent or even prostrate, 20–60 cm. long; racemes slenderer and shorter than in *Echinochloa crusgalli* because of fewer spikelets along one side of the racemes and the racemes are 1–2 cm. long. In certain forms, the leaves have reddish-purple bands across them at short intervals, an

Fig. 82. *Echinochloa colonum;* racemes, a spikelet and a portion of a striped leaf

Fig. 83. *Paspalum distichum;* inflorescence and a spikelet

extraordinary feature since the blades of most grasses are unmarked.

Naturalized at low elevations throughout the state wherever there is sufficient summer moisture. Commonly it is a weed of croplands, pastures, or along irrigation ditches, roads, and occasionally in cities and towns.

58. Genus *Paspalum* Paspalum

Perennials either densely tufted or with creeping stems, commonly broad, hairy to glabrous; ligule a membrane; inflorescence consisting of usually 2 or more racemes, if paired then these more or less terminal on the culm, if several then these scattered along the upper part of the culm; racemes of usually numerous spikelets placed along one side of a flat rachis; spikelets planoconvex, the rounded side next to the rachis.

1. Racemes 2 or 3 on the culm..............1. *P. distichum*
1a. Racemes more than 3 on the culm........2. *P. dilatatum*

1. *Paspalum distichum* L. (pl. 7h; fig. 83) Knotgrass

Perennial by creeping stems above and below ground; stolons thick, leafy, extensively creeping over the surface forming a dense mat; erect stems 20–60 cm. tall; racemes 2–7 cm. long, ordinarily 2 and standing as a V at the apex of the stem, but sometimes, in robust plants, a third one occurs below the V; spikelets sparsely hairy, elliptic, 2.5–3.5 mm. long.

Native. At low elevations throughout California, primarily in standing water or in moist bottomlands that are periodically flooded. When growing in water the plant develops an extensive rhizome system as well as numerous erect flowering stems; the upper part of the stem and the racemes are above the water level. Following recession of the water long stolons are produced, causing a rapid spread of the plant over a wide area. Because of this type of growth, Knotgrass becomes valuable in preventing erosion of ditch and stream banks, but in small ditches it tends to choke up the channel. Sometimes Knotgrass becomes a serious weed in turfgrass and is difficult to eradicate. It is a valuable forage for grazing animals, and good stands of it frequently develop about ponds or reservoirs or along creeks in the valley and foothill range areas.

2. *Paspalum dilatatum* (fig. 84) Dallisgrass

Densely tufted perennial often forming large clumps, flowering primarily in the summer period; culms stout 50–150 cm. tall; foliage with some long hairs at the base of the blades and rather densely hairy on the basal sheaths; 5–10 racemes, scattered along the upper part of the stem; spikelets ovate, 3–3.5 mm. long, with long silky hairs along the margin and shorter hairs over the surface.

Fig. 84. *Paspalum dilatatum*; a basal portion of the plant, the inflorescence, a spikelet and a portion of a leaf

Introduced from South America. Now naturalized and widely used as a hay and pasture species. As pasture the grass should be regularly grazed or mowed to prevent development of large mounds of foliage and stems. Dallisgrass has fallen into disfavor in California because of its weedy character outside of irrigated pastures. The spikelets easily float away in the excess water from the pasture and seedlings eventually become established along ditches. Dense stands of the grass greatly impede the flow of water in irrigation in drainage channels. Dallisgrass is frequent about older pastures, meadows in the foothills, seeps on hillslopes, and particularly along roads. It occurs where there is sufficient summer moisture primarily below about 3000 feet throughout much of California.

59. Genus *Pennisetum* Pennisetum

1. Stoloniferous and rhizomatous plants...1. *P. clandestinum*
1a. Densely tufted plants.

2. Panicle whitish, oblong.............2. *P. villosum*
2a. Panicle purplish, elongate...........3. *P. setaceum*

1. *Pennisetum clandestinum* (fig. 85) Kikuyugrass

Low mat-forming, long-lived perennial with creeping rhizomes and stout stout stolons; foliage hairy, the blades rather elongate on the vegetative (sterile) stems; fertile stems with stiff, short blades bearing 2–4 spikelets in the upper sheath, stigmas long-exserted. The specific name, *clandestinum*, refers to these quite inconspicuous or hidden spikelets.

Introduced from tropical Africa to check soil erosion along ditchbanks, arroyos and slopes in southern California. Though quite effective in developing a mat its aggressive and weedy character soon became a problem in citrus groves, orchards, and other crop lands. Stringent controls and eradication programs have removed much of it from the valuable crop areas.

In southern California and along the coast to the San Francisco Bay area the grass is sometimes useful as a turfgrass often producing a quick cover and growing in areas where other grasses would do poorly. Bluegrass and Bentgrass lawns along the coast have been invaded by, and sometimes replaced by, Kikuyugrass.

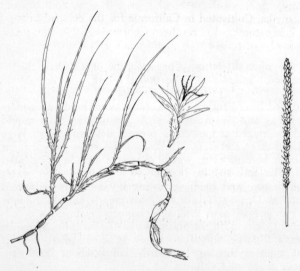

Fig. 85. *Pennisetum clandestinum* showing the habit of the plant and a flowering shoot with exposed stigmas
Fig. 86. The spikelike panicle of *Setaria geniculata*

2. *Pennisetum villosum* (pl. 8a) Feathertop

Densely tufted perennial; culms 20–80 cm. tall, hairy below the panicle; foliage more or less hairy, rather densely so about the throat of the sheath and the sheath margins; panicles oblong in outline, spikelike, 5–10 cm. long, very bristly, the bristles long, whitish and the inner ones plumose; spikelets 8–14 mm. long, solitary or in clusters of 2–4, surrounded by numerous bristles which form sort of an involucre, the whole involucre with the spikelets above falling as a unit from the densely hairy rachis.

Introduced from middle Arabia and Ethiopia. It is commonly cultivated for the attractive bristly panicles. The grass is tolerant of drought and has occasionally become naturalized.

3. *Pennisetum setaceum* (pl. 8b) Fountaingrass

Densely tufted perennial; culms 30–120 cm. tall; sheaths ciliate with conspicuous tufts of white hairs on the collar; panicles very bristly, spikelike, tapering at the apex, 15–20 cm. long, commonly violet-colored though sometimes whitish; spikelets 4.5–6.5 mm., long, solitary or in clusters with as many as 6, surrounded by ordinarily brownish bristles, the inner ones plumose, the bristles forming an involucre; the involucre with spikelets above falling away as a unit from the hairy rachis.

Introduced from western Asia, Egypt, Ethiopia, Somaliland, and Tanganyika. Cultivated in California for the colorful bristly panicles. It is naturalized in southern California and is especially common on the disturbed soils along roads and freeways.

60. Genus *Setaria* Bristlegrass

Annuals or tufted perennials; culms branching from the base and sometimes from the culm nodes as well; ligule a membrane fringed with hairs; panicles in most California species dense, spikelike, cylindrical, the branches and pedicels very short appearing as stubs with cup-shaped apices these quite evident after the fall of the spikelets; bristles below spikelets persistent, few to numerous, surrounding and mostly exceeding the length of the spikelets; lower glume well developed, broad, 3–5–nerved, about half as long as the spikelet; fertile lemma becoming hardened, smooth or the surface roughened.

Introduced from tropical regions or some from the Old World temperate zone. The species included here are summer weeds in or around pastures, orchards, vineyards, other crops, lawns and ornamental plantings. Good stands often develop in ditches or along roads or along streams. Bristlegrasses occur at low to sometimes medium elevations throughout California. The bristly panicles are used in dried flower arrangements even though the spikelets readily fall away.

1. Perennial..............................1. S. *geniculata*
1a. Annual.
    2. Bristles downwardly barbed; panicle lobed or interrupted..........................2. S. *verticillata*
    2a. Bristles upwardly barbed; panicle not lobed or interrupted.
        3. Sheath margin ciliate..............3. S. *viridis*
        3a. Sheath margin without hairs....4. S. *glauca*

1. *Setaria geniculata* (fig. 86) Knotroot Bristlegrass
Tufted perennial forming tough, knotty crowns; culms leafy and

wiry, 35–90 cm. tall, often reddish at the base; foliage glabrous except for a few long hairs above the ligule; panicle mostly pale green, 2.5–8 cm. long, 2.5–8 mm. wide, bristles yellowish, as long as to twice the length of the spikelets; spikelets 2–3 mm. long; fertile lemma transversely wrinkled.

2. *Setaria verticillata* (fig. 87) Hooked Bristlegrass
Annual; culms 25–90 cm. tall, often branching and rooting from the nodes above the base; foliage sparsely hairy to glabrous; panicle 4–10 cm. long, 5–10 mm. wide, at first cylindrical but becoming more or less lobed or interrupted in the lower part; bristles flexuous, purplish-tipped and peculiarly retrorsely barbed (this feature immediately identifies this Bristlegrass species); spikelets 2–2.5 mm. long; fertile lemma maturing brownish and minutely wrinkled across the surface.

Frequently the panicles come in contact and adhere to each other facilitating easy identification of this species. Sometimes other plant materials and insects are caught among the bristles.

3. *Setaria viridis* Green Bristlegrass
Annual; culms erect or geniculate, 20–80 cm. tall; sheaths densely short-hairy along the margin, the foliage otherwise glabrous; panicle 2–10 (–14) cm. long and 5–10 mm. wide; bristles pale yellow to purplish, 2–4 times the length of the spikelet; spikelets pale green or purplish, 2–2.5 mm. long; fertile lemma pale green to tan-colored, very finely wrinkled across the surface.

Fig. 87. *Setaria verticillata*; a, habit; b, panicle branch, bristles and spikelet

The closely related *Setaria italica*, Foxtail or Italian Millet, is commonly cultivated as a quick-growing cereal in many parts of the world, especially in Asia. The spikelets are a common component of bird seed mixes or the mature panicles are packaged and sold for pet bird use. Occasionally the plant volunteers and would be expected in some areas of California. Sometimes a small acreage may be grown for commercial purposes. This millet differs from *Setaria viridis* in the larger inflorescence (which is commonly over 10 cm. long and 2–3 cm. wide), often lobed panicle, and in the perfectly smooth fertile lemma.

4. *Setaria glauca* (pl. 8c; fig. 88) Yellow Bristlegrass

Annual with erect or geniculate culms 15–75 cm. tall, the young shoots strongly flattened; sheaths strongly keeled, loose on the culm, glabrous; blades flat or folded, sparsely long-hairy on the upper surface about or above the ligule; panicle 5–10 cm. long to about 1 cm. wide, bristles yellowish to brownish or reddish tipped, 2 to 3 times as long as the spikelets; spikelets pale green, 3–3.3 mm. long, strongly convex on one side; fertile lemma maturing usually a dark-brownish and is markedly wrinkled.

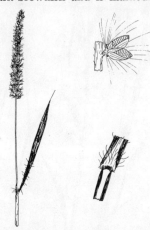

Fig. 88. *Setaria glauca* showing the inflorescence, two spikelets and associated bristles, and a portion of a leaf

61. Genus *Cenchrus* L. Sandbur

1. *Cenchrus longispinus* (figs. 89, 90) Mat Sandbur

Annual, flowering during the summer, culms 20–60 cm. long, largely decumbent, branching at the base as well as above, often forming large mats; sheaths flattened, loose on the stems; blades V-shaped or sometimes flat; racemes 4–10 cm. long with several to many spiny, hairy burs, each bur with usually 2 spikelets; spines flattish at the base, the uppermost rigid and larger than the slender lower ones.

Native. Grows well on sandy soils, particularly in the San Joaquin Valley; occasional along roadsides or disturbed soils in other parts of the state. It is a weed of vineyards, orchards and pastures. The burs are readily detached from the tip of the stems, easily become entangled in the hair of animals or clothing of man, and are quite painful when they penetrate the skin. The minute backwardly pointed barbs on the spines make removal of the bur difficult. Besides being carried about by hair or skin, the burs readily float and are carried by irrigation water sometimes for long distances. The seeds remain viable for many years and when occurring in dry fallow, remain available when the land is again brought under irrigation. Mat Sandbur is subject to eradication measures since it is a formidable weed; once started in any area it soon becomes abundant unless carefully controlled.

Fig. 89. *Cenchrus longispinus* habit
Fig. 90. *Cenchrus longispinus*; a bur greatly enlarged

62. Genus *Andropogon* Bluestem
1. *Andropogon virginicus* (pl. 8d; fig. 91)
Broomsedge; Yellowsedge Bluestem

Perennial, usually densely tufted; culms 60–100 or more cm. tall; young basal shoots strongly flattened, whitish; inflorescence quite elongate with clusters of 2–4 racemes borne on slender peduncles, the peduncles several and scattered along the stem, each associated with a bronzy leaf (or spathe); racemes white-silky, the spikelets divergent, in pairs, one represented only by a silky pedicel, the other sessile on the silky rachis and hairless but provided with a hairlike straight awn.

Introduced from the eastern half of the United States. Naturalized in California mostly at lower elevations, primarily as a weed of moist soils in old pastures, meadows, along ditches, in seeps, etc. The culms begin to elongate by late summer and the grass flowers during the fall. The flowering stems are attractive, but the racemes readily shatter into small hairy segments. In the field the seed is carried away by the wind, often for long distances.

Broomsedge is common at the edge of, and in the foothills along, the western base of the Sierra Nevada, especially along the eastern border of the Sacramento Valley.

Fig. 91. *Andropogon virginicus* showing a portion of the inflorescence and a portion of a single raceme

63. Genus *Miscanthus* Silvergrass
1. *Miscanthus sinensis* Chinese Silvergrass; Eulalia

Robust perennial with short thick rhizomes; culms 1–3 meters tall; leaves basal and well distributed on the culm; blades elongate to about 1 meter long and 1–2 cm. wide, saw-toothed on the margins; racemes numerous, silky, 10–25 cm. long, in a more or less fan-shaped arrangement about the summit of the culm; paired spikelets alike but unequally pedicelled along a slender nondisjointing rachis; callus hairs as long as or exceeding the spikelet.

Introduced from China and Japan. Occasionally cultivated as an ornamental because of its hardiness, height, and attractive inflorescence. Rarely is it spontaneous in California. The inflorescences are dramatic in dried flower arrangements and sometimes may be displayed alone for special effect.

64. Genus *Sorghum* Sorghum

Ordinarily robust annuals or sometimes perennials; blades flat often exceeding 1 cm. wide and with a broad white midrib; ligule a short membrane fringed with hairs; inflorescence a large open or contracted panicle with the spikelets organized into 1–5-jointed racemes that are scattered along the branchlets; spikelets ing stamens, the terminal spikelet associated with 2

Fig. 92. *Sorghum halepense*; a portion of the inflorescence showing a single raceme

in pairs, the sessile one fertile, the pedicelled one rather well developed but sterile although sometimes producpedicelled spikelets; fertile spikelet with 2 florets, the lower reduced to a lemma, the upper grain-producing and usually awned, the awn twisted below and bent above; lemmas thin the paleas minute or absent; glumes tough, leathery.

1. Rhizomatous perennial..................1. *S. halepense*
1a. Annual
2. Spikelets lanceolate to elliptic; blades 2 cm. or less wide..
2. *S. sudanense*
2a. Spikelets ovate to globose; blades 3 cm. or more wide....
3. *S. bicolor*

1. *Sorghum halepense* (pl. 8e; fig. 92) Johnsongrass

Summer-maturing perennial with stout, whitish, extensively creeping rhizomes; coarse clumps develop from the network of rhizomes; culms 50–150 cm. tall, the leaves distributed on the culm; panicle reddish-purple to pale green, 10–30 cm. long, the short racems readily breaking up at the joints and the apices of the rachis joints cup-shaped; fertile spikelet averaging about 5 mm. long, appressed-hairy, with a readily deciduous awn; sterile spikelet 5–7 mm. long.

Introduced from the Mediterranean region. Widely naturalized in California at low elevations wherever there is sufficient moisture. The aggressive rhizomatous character of Johnsongrass makes it a formidable weed in or around croplands and control or eradication measures are frequently necessary. It is common along roads or ditches and sometimes in cities and towns. As a forage grass it is good but needs management and periodic cultivation to maintain its vigor. It cannot be heavily pastured or frequently cut for the stand would rapidly decline.

2. *Sorghum sudanense* (pl. 8f) Sudangrass

Summer or fall-maturing annual; culms 150–300 cm. tall; blades 8–20 mm. wide, 15–30 cm. long; both sterile and fertile spikelets about 6 mm. long; awn more or less persistent, 10–15 mm. long.

Introduced from Africa. Widely grown in California as a seed crop, for pasture and hay, or as a rotation crop. Occasional along roads or in or about fields where the grass had been planted.

3. *Sorghum bicolor* (*Sorghum vulgare*) (pl. 8g) Sorghum

Robust annual; culms commonly stout, glabrous or smooth, of widely differing heights, according to variety, from about 60 cm.-4 meters tall; blades 3–5 cm. wide; panicles extremely variable, depending upon the strain or variety, loose to dense, erect to nodding, the stalk of the inflorescence erect or bent downwards; fertile spikelets ovate to globose, varying as to color from yellowish to orange, red, or even blackish.

Cultivated extensively in many different strains and types, in California, mostly for grain as animal feed but sometimes as forage or as silage. In other parts of the world Sorghum is used as human food; the sweetish stems are useful for their sugar content.

65. Genus *Zea* Corn; Maize

1. *Zea mays* (pl. 8h) Corn; Maize

Tall robust annual with thick, canelike stems; sheaths overlapping with broad, flat, distichous blades; staminate spikelets arranged in spikelike racemes, these numerous in a large terminal panicle (the tassel); pistillate spikelets sessile, in pairs and in several to many lines on a thickened corky or rather woody axis (the cob), the whole enclosed by several to many large leaflike spathes and making up the *ear*, the ears arising in the axils of the leaves; long styles (silk) protrude from the summit of the ears; mature grains greatly exceeding the thin glumes.

A widely cultivated crop as food for man and domesticated animals.

THE COLLECTION OF GRASSES AND THE PRESERVATION OF SPECIMENS

Grasses are best studied by making a dried and pressed collection. Remove growing plants from the soil or, if they are in large clumps, then take only a portion of the crown with leaves, stem, and inflorescence. Place the specimen between a single page of newspaper folded in the middle, and put absorbent paper on either side. (The absorbing paper may be several pages of newspaper or a special blotting paper.) A large stack of specimens may be assembled in this manner, and the whole then weighted down. The weight may be bricks or other construction stone; books; or two boards placed on either side of the stack of specimens and secured by web-type straps that can be cinched up. A weight or pressure of 20 pounds or more is recommended to press and eventually dry the plant material. It is essential to change the water-absorbing paper on either side of the plant specimen about once a day. Most grasses are easy to dry, requiring only a few days, but very succulent grasses may require a longer drying time and much attention must be given to changing the absorbing paper.

Following drying, remove the pressed specimen, still enclosed in the single fold of newspaper, from the drying stack. It may then be stored in a cabinet or box or even mounted on stiff white paper for display. A plant specimen thus dried and pressed may be preserved and contained within a relatively small area for many years, providing it is stored in a dry place and protected from insects and rodents. Insect larvae and some adults, such as weevils, feed upon grass grains and leaves or stems and can disfigure or destroy the plant material. Rodents may nest in or chew the newspapers containing the plant specimens.

To increase the value of your California grass collection, certain information is required for each specimen. First in importance is the identity of the species, which you can probably discover through the use of the keys in this guide. Second, note where the grass grows—a series of specimens of one species with their locations gives a record of distribution of that species in California. Third, describe the habitat; this clarifies the *nature* of the distribution of a species. Fourth, note the date on which the specimen was collected; this characterizes its response to climate. Fifth, include the name of the collector of the grass specimen, even if you are the only collector; it then becomes a permanent record of who contributed knowledge of the occurrence of a species in nature. Sixth, make notes on plant height, color, and perhaps density.

Some organizational sequence is necessary for a good grass collection. An unbroken numerical sequence is the most efficient, beginning with number 1 for the first grass collected. Clearly attach the assigned number to the specimen at the time of collection. For each numbered specimen, record the descriptive data separately in a bound notebook. Include, besides the number, the identity, geographic location, habitat, date of collection, collector's initials, and other notes. Care must be taken in the assignment and maintenance of the numbers as well as of the specimens to which they apply.

A sample page from your notebook may appear like this (recording the last grass collection at the end of summer and the next collection beginning in the spring):

178. Digitaria sanguinalis. Hairy Crabgrass
   Solano County: 1 mile west of Elmira. Common along the edge of the road in gravelly soil. Plants gray-green with slight purplish tinge; stems to 12 inches high. Associated with *Panicum hillmanii*.
   BC. September 21, 1972.

179. Festuca megalura. Foxtail Fescue
    Yolo County: Davis. Common, forming dense stands along banks of the railroad; gravelly and sandy soil. Plants yellow-green, 6–10 inches high. Associated with *Bromus diandrus* and *Hordeum leporinum*.
    BC. April 15, 1973.
180.
    Sacramento County; Sacramento. Sandy soil along the bank of the Sacramento River at Tower Bridge. Plants with purple tinge, 6–8 inches high, associated with *Bromus diandrus* and *Erodium botrys*.
    BC. April 30, 1973.

The identification of a grass specimen in your collection may not be possible immediately. It can be made at any time and recorded in the notebook opposite the number to which it applies, as was done in numbers 178 and 179 above. In the case of number 180 the identification has yet to be made. The notebook is extremely important as it carries information about the plant specimen above and beyond its identity. It becomes a reference from which appropriate labels can be prepared to accompany mounted or unmounted specimens.

A collection of labeled dried or otherwise preserved plant specimens arranged in a usable and systematic order constitutes a *herbarium*. Any herbarium is a record of diversity and variability among plants. The form, structure, and arrangement of all plant parts are available there for study or display. Coloration, fleshy character, underground parts, and height are not always apparent in the specimens but such characteristics may be described on the accompanying label. Your own personal California grasses herbarium can provide a ready reference to identity, similarities, and differences among the many species. As your collection increases so will your understanding of this remarkable plant family.

Excellent grass collections are maintained in the state's large herbaria: The University of California at Berkeley; The California Academy of Sciences in San Francisco; Stanford University at Stanford; and Rancho Santa Ana Botanic Gardens in Claremont. The herbarium of the Department of Agronomy and Range Science on the Davis campus of the University of California specializes in grass collection and has on file about 30,000 specimens, many of which are Californian. As a public service, herbaria ordinarily will identify a plant for the amateur. Furthermore the staff of herbaria can offer suggestions as to how collections are made, mounting techniques, and methods of preserving specimens. Many of the larger herbaria use specimens in research as a basis for taxonomic studies, as records and vouchers for new species, and for determining distributions of known species. Very often herbarium specimens are valuable as a record of host plants for plant diseases, insects, lichens, and algae. Herbarium collections are useful in teaching, particularly when a variety of forms or the variation within a single group is to be studied, and they serve as a catalog or library of the living resources of our environment.

GLOSSARY

Acuminate: long-tapering to a sharp point

Acute: converging to a sharp point

Aggregate: collected together in groups or bunches

Annual: a plant which germinates from seed, grows, flowers, disperses its seed and then dies, all within one year

Antrorse: directed upwards or towards the tip or apex; commonly applies to the placement of barbs and hairs

Appressed: close to or flatly pressed against

Ascending: in an upward direction

Attenuate: gradually tapering; a slender apex or base of a part

Auricle: in grasses, small lobes or clawlike appendages at the base of the blades

Auriculate: provided with lobes

Awn: a bristlelike appendage, usually a continuation of the midnerve of the lemma or the glume

Axis: the main stem

Barbed: provided with usually small, hard, pointed projections

Bifid: two-lobed, usually referring to the apex of the lemma or palea

Blade: the laterally expanded portion of the leaf

Bract: a much reduced leaf

Bristle: a stiff hairlike structure

Bunchgrass: a densely tufted perennial grass; a compact cluster of stems, shoots and leaves

Callus: hardened or thickened area at the point of attachment

Canescent: grayish-white or hoary, the surface covered with fine, dense short hairs

Capillary: hairlike

Caryopsis: the grain or fruit of grasses

Ciliate: a margin lined with hairs

Cismontane: "This side of the mountains," referring to that area in California west of the Sierra Nevada crest. excluding the deserts

Cleistogamous: flowers not opening, self-fertilized

Cleistogene: a plant producing cleistogamous flowers; in grasses these often enclosed by the lower sheath as in *Danthonia* and *Muhlenbergia microsperma*. Some annual fescue species produce cleistogamous flowers in the panicle.

Collar: in grasses the outer side of the leaf at the junction of the sheath and blade, often lighter colored than the surrounding tissue

Contracted: Narrowed or shortened, in opposition to open or spreading

Crown: in grasses ordinarily applying to the persistent base of a tufted perennial; sometimes referring to the apex of a floret where there is a ring of hairs as in certain *Stipa* species

Culm: the stem of grasses, usually hollow except at the ordinarily swollen nodes

Deciduous: falling away from, as florets from the spikelets, spikelets from the pedicels, awn from the lemma, etc.

Decumbent: lying on the ground with the tip ascending

Depauperate: poorly developed, small or spindly

Diffuse: loosely branching or spreading; very open growth

Digitate: radiating from a common point

Distichous: two-ranked, as leaves on opposite sides of a stem and in the same plane

Divaricate: widely and usually stiffly spreading

Drooping: erect or spreading at the base, the ends bending downwards

Elliptic: widest about the middle, the ends narrower and rounded

Elongate: narrow and stretched out; many times as long as wide

Endemic: native to or confined naturally to a particular region; a restricted distribution

Exserted: projecting beyond

Fascicle: a close, condensed cluster

Fertile: capable of producing fruit; provided with a pistil

Filiform: long and slender; threadlike

Flexuous: having a wavy form

Floret: individual flowers bracted by a lemma and palea; florets may be bisexual, unisexual, or neuter

Fruit: a ripened ovary (pistil); in grasses the caryopsis or grain but sometimes interpreted as also including the lemma, palea, and/or other parts of the inflorescence

Geniculate: bent, like a knee

Glabrous: without hairs; should *not* be used or applied as *smooth*

Gland: a secreting protuberance or depression

Glaucous: covered with a whitish bloom or waxy coating imparting a bluish-green color

Glumes: a pair of sterile bracts at the base of the spikelet

Grain: the fruit of grasses; caryopsis

Gregarious: growing in groups or colonies

Herbaceous: not woody, dying down each year

Hispid: with stiff or bristly hairs

Hyaline: thin, transluscent or transparent

Included: said of an inflorescence enclosed at least partly by the upper leaf sheath

Inflated: blown up; bladdery

Inflorescence: the flowering portion of a plant

Internode: between two nodes

Interrupted: not continuous; with gaps
Involucre: one or more whorls of bracts below a flower cluster
Involute: with the edges rolled inwards towards the upper side
Joint: the culm node; or sometimes referring to the internode of a jointed rachis
Keel: a prominent ridge along the back resembling the bottom of a boat
Lacerate: torn along the edge and sometimes deeply so forming a fringe of fine slender segments
Lanceolate: much longer than broad, widening above the base and tapering at the apex; shaped like a lance
Lateral: on or at the side of
Leaf: in grasses, the sheath and blade
Lemma: the lower of 2 bracts surrounding the flower
Linear: long and narrow, the sides parallel or nearly so
Lobe: usually a rounded segment of an organ
Membranous: like parchment in texture
Midnerve: the central vascular (vein) strand
Midrib: the main vein or central vascular strand
Mucronate: ending abruptly in a sharp but usually short point
Nerve: the vascular strand of a glume, lemma or leaf
Neuter: without a pistil or stamens
Node: the joint of a culm
Obovate: the upper part broader than the basal; the reverse of ovate
Obtuse: rounded
Ovate, ovoid: shaped like an egg; broadest below the middle
Palea: the uppermost bract surrounding the grass flower
Panicle: a central axis provided with solitary, several, or whorls of branches that bear pedicelled flowers
Pectinate: arranged like the teeth of a comb
Pedicel: the stalk of a spikelet
Peduncle: the stalk of an inflorescence
Perennial: lasting longer than one year; growing again from a parent plant
Perfect: said of flowers with stamens and pistil
Persistent: remaining attached; not falling off
Pilose: provided with soft straight hairs
Pistillate: female; having pistils and no functional stamens
Planoconvex: flat on one side and rounded on the other
Plumose: featherlike; having fine hairs on each side
Pubescent: covered with short, soft hairs
Pyramidal: shaped like a pyramid (usually in outline)
Raceme: flowers pedicelled along a main axis
Rachilla: a short rachis, applying to the axis of a spikelet
Rachis: the main axis of an inflorescence
Retrorse: backwardly pointed

Rhizome: a creeping underground stem
Rosette: a radial arrangement of leaves close to the ground
Rudiment: an imperfectly developed organ; a vestige
Scabrous: rough to the touch like the feel of fine sandpaper; provided with small barbs or very short, stiff hairs
Scale: a small dry bract
Scarious: thin, dry, like a membrane
Serrate: with a saw-toothed margin
Sessile: directly upon an axis; without a stalk
Setaceous: bristlelike
Sheath: in grasses, the basal portion of the leaf that surrounds the stem
Smooth: without barbs or hairs
Spathe: a bract that sheaths an inflorescence
Spike: spikelets borne directly upon a main axis without intervening pedicels
Spikelet: the basic unit of a grass inflorescence consisting of 2 glumes and 1 or more flortes
Staminate: producing only stamens
Stand: the relative number of plants growing on a given area
Sterile: not grain-producing
Stipe: the stalk of a pistil or other small organ
Stolon: stems that lay upon the ground and root at the nodes or at least bend over and root at the tip
Striate: marked with longitudinal lines, grooves, or ridges
Sub-: a prefix to indicate somewhat, slightly, or to a lesser degree
Subtend: to enclose, thus attached below
Succulent: juicy, fleshy
Tawny: pale brown, dirty yellow
Throat: in grasses, the V-shaped summit of the sheath on the inner side
Tiller: the young vegetative shoot in grasses
Transverse: across
Triad: a cluster of 3, as spikelets of *Hordeum* and *Hilaria*
Truncate: cut off squarely or nearly so at the end
Tufted: a close aggregation of stems or shoots and leaves, as typical of bunchgrasses
Villous: bearing long, soft, but not matted hairs; shaggy
Web: a cluster of long hairs at the base of a floret as in certain *Poa* species
Whorl: a circular cluster of several branches at a single node of a panicle
Wing: a thin projection or border as along the keel of the glumes of *Phalaris minor*

SELECTED READINGS AND REFERENCES

Abrams, Leroy. *An Illustrated Flora of the Pacific States* Stanford University Press, 1923–1960. 4 vols., vol. 1 for grasses.

Beetle, A. A. "Distribution of the Native Grasses of California." *Hilgardia*, 1947. Vol. 17, no. 9, pp. 309–357.

Bews, J. W. *The World's Grasses, Their Differentiation, Distribution, Economics and Ecology.* Longman, Green, 1929.

Chase, Agnes. *The First Book of Grasses. The Structure of Grasses Explained for Beginners.* Smithsonian Institution, Washington, D. C., 1959.

Gould, Frank. *Grass Systematics.* McGraw-Hill, 1968.

Hitchcock, A. S. *Manual of the Grasses of the United States* 2nd ed., rev. by Agnes Chase. U. S. Dept. of Agriculture, Misc. Publ. no. 200, Feb. 1951. Nearly all species illustrated. (Now in 2 vol., Dover Publications, N. Y., 1971).

Hubbard, C. E. *Grasses.* 2nd ed. Pelican Book A 295, Penguin Books, 1968. About grasses of the British Isles some of which also occur in California; well illustrated.

Jepson, W. L. *A Manual of the Flowering Plants of California.* University of California Press, 1970.

Munz, P. A. in collaboration with David D. Keck. *A California Flora.* University of California Press, 1959.

Pohl, Richard W. *How to Know the Grasses. Picutred keys of the commonest and most important American grasses in farming, gardening, weed control, range and pastures.* W. C. Brown Co., Dubuque, Iowa, 1968.

Sampson, A. W., Agnes Chase, and D. W. Hedrick. *California Grasslands and Range Forage Plants.* California Agricultural Experiment Station, Bulletin 724, 1951.

INDEX

Aegilops triuncialis, 78
Agropyron, 89
  desertorum, 90
  intermedium, 92
  repens, 91
  spicatum, 91
    var. inerme, 91
  subsecundum, 30
  trachycaulum, 91
  trichophorum, 92
Agrostis, 103
  alba, 104
  aristiglumis, 35
  avenacea, 104
  blasdalei, 19, 35
  californica, 36
  clivicola, 35
  diegoensis, 36
  exarata, 106
  exigua, 30
  microphylla, 30
  scabra, 105
Aira, 95
  caryophyllea, 95
  elegans, 96
Alkali Sacaton, 130
Alopecurus, 34
  saccatus, 30
American Sloughgrass, 30
Ampelodesmos mauritanicus, 119
Andropogon virginicus, 159
Annual Semaphoregrass, 31
Annuals, 10
  summer, 11
  winter, 11
Anthoxanthum odoratum, 103
Aristida, 126
  adscensionis, 128
  oligantha, 126
Arrhenatherum elatius, 34
Arundo donax, 121
Auricle, 13
Avena, 93
  barbata, 94
  fatua, 94
  sativa, 95

Bamboo
  Fishpole, 120
  Giant, 120
  Golden, 120
  Hardy Timber, 120
  Yellow, 120
Barbs, 19
Barley, 78, 83
  Alkali, 83
  Foxtail, 80
  Glaucous, 82
  Hare, 82
  Meadow, 80
  Mediterranean, 83
Barnyardgrass, 151
Beckmannia syzigachne, 30
Bentgrass, 103
  Blasdale, 35
  California, 36
  Cliff, 35
  Pacific, 104
  Point Reyes, 35
  Sixweeks, 30
  Small-leaved, 30
  Spike, 106
Bermudagrass, 138
Binomial Nomenclature, 24
Blade, 12
Blue Grama, 37
Bluegrass, 57
  Annual, 58

Canada, 58
Howell, 31
Kentucky, 9, 59
Nevada, 59
Pine, 59
Bluestem, 159
   Yellowsedge, 159
Bouteloua gracilis, 37
Brachypodium distachyon, 77
Bristlegrass, 155
   Green, 156
   Hooked, 156
   Knotroot, 156
   Yellow, 158
Briza, 61
   maxima, 62
   minor, 61
Brome, 67
   Australian, 74
   California, 70
   Cheatgrass, 74
   Chinook, 72
   Downy, 74
   Foxtail, 76
   Hairy, 74
   Madrid, 76
   Mountain, 70
   Poverty, 75
   Prairie, 71
   Red, 76
   Rescue, 72
   Ripgut, 75
   Smooth, 72
   Soft, 73
   Spanish, 76
   Woodland, 72
Bromus, 67
   arenarius, 74
   carinatus, 70
   commutatus, 74
   diandrus, 75
   haenkeanus, 72
   inermis, 72
   laevipes, 72
   madritensis, 76
   marginatus, 70
   mollis, 73
      forma leiostachys, 74
   rigidus, 75
   rubens, 76
   sterilis, 75
   tectorum, 74
   unioloides, 72
   willdenowii, 71
Broomsedge, 159
Bunchgrasses, 7
Bur Clover, 32

Calamagrostis, 107
   canadensis, 107
   nutkaensis, 107
   rubescens, 30
California Bottlebrushgrass, 36
California Sweetgrass, 36
Canarygrass, 111
   Hood, 112
   Lemmon, 30
   Littleseed, 112
   Mediterranean, 112
   Reed, 30
Caryopsis, 18
Cenchrus, 158
   longispinus, 158
Chess, 67
   Soft, 73
Chloris, 137
   Showy, 137
   virgata, 137
Clovers, 33
Cockspur, 149
Collar, 13
Corn, 162
Cortaderia selloana, 122
Crabgrass, 147
   Hairy, 147
   Large, 147
   Smooth, 148
Crypsis schoenoides, 131
Culm, 12
Cupgrass, 148

Prairie, 149
Southwestern, 149
Cynodon, dactylon, 138
Cynosurus echinatus, 63

Dactylis glomerata, 60
Dallisgrass, 153
Danthonia, 122
  californica, 123
    var. americana, 123
  pilosa, 124
  unispicata, 123
Darnel, 57
Deergrass, 136
Deschampsia, 97
  caespitosa, 98
  danthonioides, 98
  elongata, 99
  holciformis, 99
Digitaria, 147
  ischaemum, 148
  sanguinalis, 147
Distichlis spicata, 9, 141
  var. nana, 142
Dogtail, 63
  Hedgehog, 63
Dogtoothgrass, 138
Dropseed, 130
  Sand, 130

Echinochloa, 149
  colonum, 151
  crusgalli, 151
Ehrharta calycina, 120
Elymus, 86
  caput-medusae, 86
  cinereus, 88
  condensatus, 88
  glaucus, 87
  triticoides, 89
Epidermal structures, 19
Eragrostis, 128
  cilianensis, 128
  diffusa, 130
  orcuttiana, 129
Eriochloa, 148

contracta, 149
gracilis, 149
Erodium, 32
Eulalia, 160

Falsebrome, 77
  Purple, 77
Feathertop, 155
Fescue, 49
  Blue, 49
  Brome, 54
  California, 50
  Foxtail, 54
  Idaho, 51
  Rattail, 54
  Red, 52
  Reed, 50
  Sixweeks, 31
  Small, 53
  Tall, 50
  Western, 51
Festuca, 49
  arundinacea, 50
  bromoides, 54
  californica, 50
  idahoensis, 51
  megalura, 54
  microstachys, 53
  myuros, 54
  occidentalis, 51
  octoflora, 31
  ovina var. glauca, 49
  rubra, 52
Filarees, 32
Fingergrass, 137
  Feather, 137
Florets, 16
Flower, 17
Fluffgrass, 37
Fountaingrass, 155

Galletagrass, 140
  Big, 140
  Woolly, 140
Gástridium ventricosum, 109
Glands, 20

Glumes, 16
Glyceria, 66
  elata, 67
  pauciflora, 66
Goatgrass, 78
  Barb, 78
Goldentop, 62
Grass flower, 17, 18
Grass plant, 8
Grasses
  collection of, 163
  description of, 49
  distribution, 29
    coastal area, 35
    crops, 38
    desert area, 36
    montane area, 33
    valley-foothill area, 29
  identification of, 24
  key, 39
  relationships of, 24
  specimens of, 163
  subfamilies, 27
  taxonomy, 25
  tribes of, 27
  uses of, 22

Hairgrass, 95, 97
  Annual, 98
  Elegant, 96
  Pacific, 99
  Silver, 95
  Slender, 99
  Tufted, 98
Hairs, 19
Hardinggrass, 112
Heleochloa schoenoides, 131
Herbarium, 165
Hierochloe occidentalis, 36
Hilaria, 140
  rigida, 140
Holcus lanatus, 96
Hordeum, 78
  brachyantherum, 80
  depressum, 83
  geniculatum, 83
  glaucum, 82
  jubatum, 80
  leporinum, 82
  vulgare, 83
Hystrix californica, 36

Inflorescence, 6, 13
Internode, 12

Johnsongrass, 161

Kikuyugrass, 154
Knotgrass, 152
Koeleria, 99
  cristata, 99

Lamarckia aurea, 62
Leaves, 12
Leersia oryzoides, 30
Lemma, 17
Leptochloa, 139
  fascicularis, 140
  uninervia, 140
Ligule, 13
Lodicules, 18
Lolium, 55
  multiflorum, 55
  perenne, 55
  temulentum, 57
Lovegrass, 128
  Orcutt, 129
  Spreading, 130

Maize, 162
Mannagrass, 66
  Tall, 67
  Weak, 66
Medicago polymorpha, 32
Medusahead, 86
Melic, 64
  California, 65
  Coastrange, 65
  Torrey, 66
Melica, 64
  californica, 65
  imperfecta, 65

torreyana, 66
Millet
  Foxtail, 57
  Italian, 157
Miscanthus sinensis, 160
Muhlenbergia, 133
  filiformis, 136
  microsperma, 134
  richardsonis, 137
  rigens, 136
Muhly, 133
  Littleseed, 134
  Mat, 137
  Pullup, 136
  Slender, 136

Needle-and-thread, 115
Neostapfia colusana, 143
Nitgrass, 109
Node, 12
Nuttall Alkaligrass, 30

Oat, 93, 95
  Slender, 94
  Wild, 94
Oatgrass, 122
  Americana, 123
  California, 123
  Hairy, 124
  Onespike, 123
  Tall, 34
Oniongrass, 64
Orchardgrass, 60
Orcuttia, 31
Oryza sativa, 120
Oryzopsis, 117
  hymenoides, 119
  miliacea, 118

Pacific Foxtail, 30
Pampasgrass, 122
Panicle, 13
Panicum, 143
  capillare, 145
  Desert, 37
  dichotomiflorum, 146
  Fall, 146
  Hillman, 146
  hillmanii, 146
  lanuginosum var.
    fasciculatum, 144
  Pacific, 144
  pacificum, 144
  urvilleanum, 37
Paspalum, 152
  dilatatum, 153
  distichum, 152
Pennisetum, 153
  clandestinum, 154
  setaceum, 155
  villosum, 155
Perennials, 6
Phalaris, 111
  arundinacea, 30
  lemmonii, 30
  minor, 112
  paradoxa, 112
  tuberosa var. stenoptera, 112
Phleum, 110
  alpinum, 111
  pratense, 111
Phragmites, 121
  australis, 121
  communis, 121
Phyllostachys, 119
  bambusoides, 120
    var. aurea, 120
    var. bambusoides, 120
Pinegrass, 30
Plant key, 24
Pleuropogon californicus, 31
Poa, 57
  annua, 58
  compressa, 58
  howellii, 31
  nevadensis, 59
  pratensis, 9, 59
  scabrella, 59
Polypogon, 108
  australis, 109
  Chilean, 109

maritimus, 109
Mediterranean, 109
monspeliensis, 108
Prairie Junegrass, 99
Prairie Wedgegrass, 30
Pricklegrass, 131
Puccinellia nuttalliana, 30

Quackgrass, 91
Quakinggrass, 61
  Big, 62
  Little, 61

Rabbitfootgrass, 108
Raceme, 14
Rachilla, 7
Redtop, 104
Reed, 121
  Common, 121
  Giant, 121
Reedgrass, 107
  Bluejoint, 107
  Pacific, 107
Rhizomes, 8
Rice, 120
  Jungle, 151
Rice Cutgrass, 30
Ricegrass, 117
  Indian, 119
Rye, 93
Ryegrass, 55
  Annual, 55
  Darnel, 57
  English, 55
  Italian, 55
  Perennial, 55

Saltgrass, 9, 141
Sandbur, 158
  Mat, 158
Schismus, 124
  Arabian, 125
  arabicus, 125
  barbatus, 125
  Mediterranean, 125
Scribneria bolanderi, 31

Secale cereale, 93
Seed dispersal, 20
Setaria, 155
  geniculata, 156
  glauca, 158
  italica, 157
  verticillata, 156
  viridis, 156
Sheath, 12
Silvergrass, 160
  Chinese, 160
Sitanion, 84
  hystrix, 84
  jubatum, 85
Smilo, 118
Sod-forming grasses, 7
Sorghum, 160, 162
  bicolor, 162
  halepense, 161
  sudanense, 162
  vulgare, 162
Sphenopholis obtusata, 30
Spikelets, 15
Spikes, 15
Sporobolus, 130
  airoides, 130
  cryptandrus, 130
Sprangletop, 139
  Bearded, 140
  Mexican, 140
Squirreltail, 84
  Big, 85
  Bottlebrush, 84
Stems, 12
Stinkgrass, 128
Stipa, 113
  California, 117
  californica, 117
  cernua, 115
  Columbia, 117
  columbiana, 117
  comata, 115
    var. intermedia, 116
  coronata, 36
  Crested, 36
  Desert, 116

Elmer, 117
elmeri, 117
Foothill, 115
Lemmon, 117
lemmonii, 117
lepida, 115
Nodding, 115
occidentalis, 116
pulchra, 115
Purple, 115
speciosa, 116
Western, 116
Stolons, 8
Subfamilies, 27
Sudangrass, 162
Swampgrass, 131

Taeniatherum asperum, 86
Thingrass, 36
Threeawn, 126
  Prairie, 126
  Sixweeks, 128
Ticklegrass, 105
Tillers, 6
Timothy, 111
  Alpine, 111
Tribes, 26, 27
Tridens pulchellus, 37
Trifolium, 33
Trisetum, 100
  canescens, 102
  spicatum, 102
  Spike, 102
  Tall, 102
Triticum aestivum, 78

Veldtgrass, 120
Velvetgrass, 96
  Common, 96
Vernalgrass, 103
  Sweet, 103

Watergrass, 151
Wheat, 78
Wheatgrass, 89
  Bearded, 30
  Bluebunch, 91
  Desert, 90
  Intermediate, 92
  Pubescent, 92
  Slender, 91
  Stiffhair, 92
Wildrye, 86
  Ashy, 88
  Basin, 88
  Beardless, 89
  Blue, 87
  Creeping, 89
  Giant, 88
Windmillgrass, 137
Witchgrass, **Common, 145**

Zea mays, 162